The Battle After the War: DBG

Mark LeClair

Cover Art: SelfPubBookCovers.com/ Viergacht

The Battle After the War: DBG

CONTENTS

WHO IS DAVID BERNARD GOODRIDGE?

<u>CHAPTER 1</u>
WHO IS DAVID BERNARD GOODRIDGE?

Who is David Bernard Goodridge and what does he have to do with anything? Who am I, for that matter? My name is Timothy Fergus and I go by Tim, or Timmy...actually, in the military, I've been given the nicknames: Squirt, Pip-squeak, Stick-Figure and many more unflattering names. I am rather small-framed for my size, but I've always been small, so I guess what I'm saying is: I am used to being the recipient of some rather unflattering nicknames.

But I digress, who am I? I'm a veteran who was a firefighter in the Navy, Damage Controlman to be exact. Most who have heard me say I was a firefighter in the Navy have been surprised and replied with: "how can someone so small fight fires or be on a fire team?" and I suppose that's a justifiable assessment upon first glance. My inner drive and determination have usually outworked societal norms or guidelines. My mother taught me to dig deep and fight to make forward progression. She will always be my idol.

My mother has a key role in this story, because she was the main nurse for Bernie. Bernie is what David Bernard Goodridge preferred to be called. I'll never forget Bernie for the remainder of my life. My mother took care of Bernie in a Veterans Hospital, before he passed away. Bernie lived at the Veterans Hospital, where my mom, Samantha, worked as his daytime nurse. He was her favorite patient. Bernie was quite the character and my mother used to talk about him every time we spoke.

Why is this important? Because Bernie gave me information so valuable, that it not only saved me from a lifetime of problems and aggravation, but because of his efforts and willingness to provide me with the lessons from his experience, it has helped so many more as well. This couldn't have been possible without his help and my mother's prompting. I owe everything to Bernie and my mom, and this is a tribute to them in the hopes it will help or save others who are in similar positions or situations.

So, who is/was Bernie?

Bernie was an extraordinary man who spent almost twenty years in the Navy. He first joined to be a firefighter and his first ship provided him the mentorship and experience to be a great firefighter. He had great mentors while on his first, and only ship. They taught him many life lessons, according to him, but most importantly it helped form him into the man he was proud to be. He made no excuses and took ownership of his life.

Bernie had made rank while on his first ship and left as an E-5, or Petty Officer Second Class. He had to leave the ship because his wife was attached to the ship as well, but was temporarily assigned to shore duty while she was pregnant. She was brought back to the ship and he began his tour at a shore command. It wasn't as exciting for Bernie, as his day was repetitive and, as he put it, 'a yawn'. He yearned for excitement, but couldn't find it at that shore command.

He began to research his options and possibilities. At last, he found something!

Bernie stumbled across a Naval publication, which was published each month and listed opportunities in other fields within the Navy. While scanning this publication, something caught his eye: Special Warfare Combatant Craft Crewman, or SWCC operator. Driving jet boats and conducting maritime missions with Navy SEALs seemed to brighten his spirits and renew his spark for adventure. He read the requirements, the screenings and fitness parameters and dedicated his time each morning and afternoon, to become more fit.

Each morning Bernie would arrive quite early to work, before anyone else was there. He would do his workouts in the field across from the parking lot of the facility where he worked, and he would put in about an hour of exercise prior to the start of his workday. At lunch, he would run, or run and swim, and then eat lunch before his lunchtime ended. Then, after returning home from work, he would run almost five miles. He did this for six days each week and rested on the seventh.

Bernie failed his first screening at the special warfare Team and, although a failure might extinguish the motivation of many, it didn't seem to slow down his drive to become a Special Warfare, or SPECWAR operator. He worked out harder. He put on almost thirty pounds of muscle, lean muscle from all that running and swimming, and took the test again and passed. He put his package together and willingly ended his shore duty requirements early, so he could attend the special warfare training out in Coronado, California.

Bernie was singled out quickly out at Coronado, because he was so small. His instructors pulled him out of the class one day while the class was being exercised out on their slab (which is a different way of saying they were getting mashed by the instructors). Not sure what he got caught doing, he told me he played dumb as the instructors told him to go ahead and leave. As I remember it, his conversation went as followed:

Instructor: You can go ahead and leave.

Bernie: I'm sorry, what?

Instructor: Go ahead and take off.

Bernie: I don't understand.

Instructor: You don't fit in!

Bernie: What?

Instructor: Look around, you don't fit in.

Bernie: Oh! Oh, okay, I will tell you what...I'll leave when you all carry me out in a bag. How about that?

Instructor: Really? Okay Stick-figure, Stickman! How about this, how about any time one of the other students messes up, we beat you for it?

Bernie: I wouldn't have it any other way!

Instructor: Get back to the class! Stickman!

Bernie: Hooyah!

Bernie told me that they used the moniker Stickman as an insult, at first that is. Hooyah is a phrase the students and operators yell back as a confirmation that they understood the command given. It's a motivating shout out, which does motivate, especially when an entire class yells it out in unison. From what Bernie told me, his class made a lot of mistakes, so he did a lot of pushups, bear-crawls and sugar cookies! Sugar cookie was when he had to run down to the beach, jump into the ocean and get soaked, then exit the surf and roll around in the sand until he was covered in sand and looked like a sugar cookie.

He spoke about his time at Coronado with excitement and enjoyment. He told me many times that that was one of the highlights of his life. He gushed about the men he trained with and the conflict of personalities but the necessary comradery that was required to succeed as a team. I believe that to be the moral of most classes within Special Warfare that involves a group of individuals: to become a team regardless of the inner conflict or petty fights that occur. He loved his teammates.

Bernie told me he was messed with the entire time he was in training, so he played their game. He took pride in the fact that they wouldn't break him, regardless what they did to him, he wasn't going to quit. He was proud of the fact the instructors started to take notice of him swimming his swim buddy in and the many times he helped carry one of the bigger classmates through the runs in the soft sand. Bernie would smile and tell me that the instructors knew and still beat him down, just not as aggressively. He said the game got easier.

Bernie graduated and went to his next command, where he began to learn from the seasoned guys, how to do the job properly and everything involved. He mentioned it was hard work, but it was hard work that was required for the missions to be successful, regardless if the SEALs hated them or not. When I asked why the SEALs would hate them, he would wink and tell me: "It doesn't matter where you work, what you wear on your chest or what your title is, there will be good people and there will be bad people".

From the stories that Bernie had told me, amidst the bounty of valuable information, it became abundantly clear that Bernie loved the comradery the most. He loved how difficult and intricate the job was and how, each day, he had a group of people he called brothers that he knew he could depend on. I joined the Navy and became a firefighter because I really didn't know what I wanted to do, but I knew I needed to be challenged more. For a second, I had thought about going into Special Warfare, but thought I didn't have the build or strength to do so.

Bernie had joined the Navy early in life, to make a better life for him and his fiancé because he believed he couldn't support a family living in Boston. I was in awe of my mother's stories of Bernie, especially the stories of his injuries and what he did when he was on active duty. My mother explained that he was able to remain at his Team after his injuries, which he loved to do. She became sad when she explained that Bernie initiated the discharge he received because a handful of his teammates were under the spotlight and he wanted them to have the chance to get to retirement. He knew his time was coming to a close, despite loving his job and doing it well, he knew it was time.

I felt his story was worthy of telling, despite it being a tad strange. The information he imparted on me was filled with lessons learned and what not to do if faced with similar situations. I guess one could say Bernie's story is a cautionary tale, but one that he felt important to tell me, so I could pass it on in the hopes of saving other veterans from experiencing the hell that he had to endure. He formed the story using specific wording to avoid being heard from any of the staff in his hospital. Bernie was dying, he knew this, but didn't need any more hassles from any of the staff.

Bernie did not have a million friends, he was gruff at times, but he always told it like it was and didn't sugar coat anything. His experience is his own and his decision to speak with me, to answer any questions and fill me in on how to avoid going through what he went through, speaks volumes about his character and his love for his fellow veteran. Seems Bernie held some things close to his chest to protect others, especially my mom, which elevates his status well beyond many who believe to be the standard, when it comes to ethics and morals.

Some may not have liked Bernie because he was rough around the edges, but to me and to my mom when she was still alive, Bernie was someone to appreciate and listen to. I originally met Bernie because I had gotten injured fighting a fire on base and the roof collapsed on me. The incident caused a lot of damage to my back, neck and most of my body, and my mother recommended I speak with Bernie. I was to attend a Physical Evaluation Board (PEB), which is also called a medical review board, to determine if my disabilities were to the extent where I couldn't serve in my-then position. If I couldn't do that or any other job in the Navy, the military needed to discharge me. If I was able to do another job, they could determine that I change rates (jobs) and continue on active duty.

As a favor to my mother, Bernie agreed to speak with me and provide me with helpful information on what to expect, how to approach certain things, how to organize the information, how to fight for things, where to look, how to handle myself, who to speak with, how to best be prepared for negative results and so forth. It was an abundance of information that I felt should be available to everyone that serves in the military. His stories were incredible and experiences were unbelievable at times. His health was failing towards the end of my visit, but I wanted his story to be shared with the world, as Bernie helped shape my world and my role in it.

Throughout my experience with Bernie, my mother became noticeably upset, but never wanted to open up to me about what. I think it had something to do with Bernie and the bruises I spotted on him on a couple of my visits. My mother was a caregiver that invested all of her heart to their care, especially within the veteran community and that was why all of the patients loved her. She was Bernie's caregiver and he was in good hands, my mother made sure he knew that.

Bernie will always be the hero I look up and strive to be, as he instilled into me self-pride and a respect for all. His big contribution, which he joked about often, was to not sweat the little things...and it was all little things! Most will never know what Bernie had done to protect others and to ensure others never had to experience what he went through, and much of that I will never know either, but I have a good idea. He wanted me to be precise in my notes and to review each day's notes so if I had questions that evening, I could ask about it the following day. At least Bernie did not leave this existence alone, my mom was there to escort him.

My injuries left me with a large list of disabilities, such as: nerve damage throughout my body, chronic back and neck pain, torn tendons, tendonitis, fibromyalgia, memory loss, hearing loss and damage, breathing issues and much, much more. I experienced much of what Bernie told me about and, because of his guidance and information, I was able to approach each hurdle with a plan. My notes and organization guided me efficiently and aggravation-free, for the most part.

I will end this chapter with the events that ended my career in the Navy and I hope this story reaches you on some level, and saves you from any future headaches or aggravation. It's sad when few can be trusted with information passed on, because that was all our ancestors had many eons ago, the spoken word and stories. Today, I believe we have gotten lazy and dependent on technology, and we somehow forgot that technology can break on us and let us down. There's a reason Bernie's guidance is repetitive in parts, and after my experience, I can tell you that that repetitiveness was and is still appreciated.

My story is not nearly as exciting as Bernie's and my career was cut short because of the injuries I incurred while fighting a building fire on base. It was one of the old warehouses on the pier that caught fire and it was assumed to be a chemical, fuel fire that was out of control. My firehouse was the first to respond and we were there rather quickly. We were well trained and prepared for any fire on base and we trained often for various casualty events. This was another, typical building fire on base...or so I thought.

To avoid droning on about the events of that faithful day when my career, and almost my life, came to an end, I will quickly encapsulate what occurred. The building was old, the roof was weak and the fire made it weaker. I was inside the building with the fire team, manning a hose as usual, when a loud snap was heard from above. The nozzleman looked up, pointed and took off, dropping the hose on the ground. When the nozzle hit the ground, the plunger moved and the water was released, thus sending the hose into a wild dance.

The others looked up and took off behind the nozzleman, but I was transfixed on the wild hose ahead of me and the raging fire all around. I ran for the nozzle as a louder crack above me was heard. From what everyone has told me, that was the point where the roof collapsed as I was darting after the nozzle to secure it. The beam came down and clipped me, which sent me flying across the floor and into a stack of wooden crates. I don't remember much after that.

From what I was told, one of the hose handlers spotted me flying across the floor after the loud crash and ran through the flames to drag me out of the building. Others saw that hose handler and followed. Together they removed my limp, passed-out body from the building. I only remember waking up and seeing worried faces around me. I then proceeded to pass out again for a couple of days. In the end, my career was over and my body wrecked.

In the hospital, on that bed all day long, every day, I had time to think long and hard about the future of my life. If I was in this much pain now, in my early twenties, how much pain am I going to be in, in ten, twenty, thirty years? That's a dismal contemplation. I spent enough time in the hospital and witnessed enough people who spent their days in pain, and most of them were so miserable they begged for life to end. Their screams echoed through the darkened hallways on most nights. It was disturbing.

My thought was one of reflection and determination. I was still young and despite not being able to finish my military career as I had originally planned, life was not over and I still had all of my limbs attached and operable. I wasn't going to pity myself, nor allow anyone else to pity me. It was time for me to do what I needed to, to become stronger and get my discharge papers out of that place. For the record, you CAN get in trouble/yelled at for overdoing your physical therapy, apparently because it could cause more injury or damage. Who knew?

I finally got discharged and went on convalescent leave for two weeks to my mother's house. The location doesn't matter, just the fact that the universe aligned perfectly for me to meet Bernie and gain a lifetime worth of information, in only a few days. Little did I know how impactful that visit was going to be, and to be honest, I thought my mother was building Bernie up entirely too much. I thought she had a mom-crush. I was wrong, he was a great man.

Although his methods for putting information out was odd to me at first, today I absolutely understand the need for it. Bernie was cautious and only wanted me to get as much information as possible before I had to go back to base. I will never forget the first visit, nor the final one. Bernie did have a terminal illness, which my mother prepared me for prior to my arrival home, but his energy was so vibrant that I wasn't so sure he was going to die. He kept his strength up the entire time I was there, some days were better than others, but overall, he was a joy to get to know and befriend.

This was our meeting.

MEDICAL REVIEW BOARD

CHAPTER 2
MEDICAL REVIEW BOARD

Since the building collapsed on me while fighting a fire, my injuries seemed to unravel and grow. The damage was more extensive than I originally thought. I thought (more like I hoped) it was just my spine that contained all of the damage and problem areas, but I was wrong. The list became extensive.

My back was busted up, I had nerve damage, since I was knocked unconscious, I had traumatic brain injuries (TBI), most of my limbs were numb, nerve damage caused a pupil to get large (dilate) while the other pupil remained normal (which looked weird), breathing issues, my memory was got worse, I had tender points throughout my body(fibromyalgia), my ankle and leg was broken (I had to enjoy a walking boot for quite a while and crutches) and I had issues with my tendons. I thought the ringing in my ears would go away and my hearing would get better, but that didn't happen either, which was an added bonus!

I was told I had to appear before a medical review board (PEB) to determine if I could remain in the Navy and as soon as I told my mother that news, she immediately asked Bernie if I could pick his brain for information that could help me through it. Of course, he said yes, but I don't think I was prepared for the information he was going to provide me. I was told by others that the PEB was a formality and if they decided I was to be discharged, the Kingdom would take care of me. I trusted those who informed me of what to expect if I were discharged. I was comfortable and naïve that I was going to be taken care of properly. I...was...wrong!

Since my mother had to work at the hospital anyway, it made sense that I went with her to visit Bernie. My mother had prepped me as we approached the hospital, that Bernie was terminal, but that I should avoid bringing that up. I was baffled that my mother would think that I would have nonchalantly brought up the fact that he was dying. Baffled! I think she was nervous for me to meet Bernie and kept saying he was very different and not liked by everyone. I asked my mother if she liked him and when she said she did, I reminded her that that was good enough for me.

The section of the hospital, rather a separate building from the main hospital, was the older buildings that the new buildings were built around. These buildings didn't look new and shiny, as the new building looked, and it was apparent that the older buildings were not on the priority list for upgrades or repairs. These older buildings were the locations where the terminally ill veterans lived. One look and I could quickly gain my opinion of the neglect, and my aggravation slowly simmered inside.

Upon arriving on my mother's floor, which was Bernie's floor also, my mother quickly reminded me to not cause trouble and be nice to Bernie. Again, I was amazed at her for thinking I would be anything but thankful for his help. She shooed me away as she darted behind the nursing station to gather up her needed items prior to her shift. I stood for a moment and assessed the state of the area I was in.

A light flickered in the hallway, the tile flooring had chunks missing from them and the wall paint was extremely stained and faded.

I slowly hobbled down the hall towards Bernie's room. I attempted to be quiet as I traveled down this hall, but the thud of my walking cast combined with the clink-clink of each crutch as it landed on the crumbling tiles made it a symphony of noise as walked. The Velcro straps on my cast crunched as I went through my steps, which echoed down the hall. I laughed at the noise I was trying not to make and stopped at Bernie's door. His nameplate had his initials in big, bold letters: DBG.

I stood for a moment to collect my thoughts when my mother whispered loudly from the nurses station: "Go on! Knock and step in! Go!". I quickly shushed her as comically as possible and dramatically tapped on his heavy, wooden door before turning the handle and slowly opening it. I stood for a moment and squinted as I looked into the dark, depressing room and was immediately curious about why it was so dark. I was afraid I was going to wake Bernie. From inside the room, a deep gruff-like voice barked out: "In or out, but you ain't a door!" which was my first interaction with Bernie. I smiled and slowly hobbled into the room as my eyes adjusted to the darkness.

"Mr. Goodridge, would..."

"Bernie! Call me Bernie"

"Okay, Bernie, would you like me to turn on a light"

"I would not"

"Okay, sorry. My name is Tim. I'm Samantha's son"

"I've been expecting you, Timothy"

"You can call me Tim, sir"

"Don't call me sir, son. I worked for a living. Call me Bernie"

"Okay Bernie, I'm going to sit down if that's okay"

"How's your leg, son?"

"It's healing"

"I'm sure it is. You're gonna be a walking barometer for the rest of your life, son"

I laughed but I knew he was right. I grew up with my grandfather accurately predicting the rain and he was always right. He had a bad leg and a bum foot and when they began to ache, he would tell me that the rain or the snow was on its way, and every time he was correct. I hobbled cautiously in the darkness as my eyes adjusted. At the foot of the bed was a recliner and that was my destination. I pulled off my backpack, dropped it next to the chair and sat down.

"Well Tim, how's your back doing? Your mom says you're healing well"

"I am, Bernie, thank you. I ache most of the day"

"I wish I could tell you it'll get better, son, but it's not going to"

"I know, but thanks. Any chance I could get a little light over here so I can take notes?"

"Lift the shades a little. The sun will brighten the room. I have vision issues and the light bothers my eyes. If the light isn't enough when you lift the shades a little, you can take the floor lamp next to the sink and bring to your chair. There's an outlet around there somewhere."

"Thanks Bernie, I'll lift the blinds a little then and see if it's enough light"

I stood up and hopped over and lifted the shade. It allowed enough light for me to see, despite only being lifted about six inches. It was bright outside and the incoming glow allowed me to see the pastel-green paint on the walls, which made the room seem brighter. This allowed me to see Bernie, somewhat. It wasn't bright enough for me to see him well, just well enough. He looked different from what I had imagined. When my mother described his personality and demeanor, I pictured Bernie to be a large, mean-looking giant with scars all over his face.

Bernie looked like a beat-up version of Santa Clause with the round face and mostly-white beard. I hopped back to the chair, turned and stood there for a moment and stared at him.

"Is that okay, Bernie?"

"It's okay with me. Is it enough light for you?"

"I think so. Thanks"

I sat down, reached into my backpack and pulled out my notebook and pen. My mother advised I bring a large notebook and a handful of pens because Bernie was a trove of information. I'm glad I listened because I never imagined the amount of writing I was in for. I prepped my area, sat back and prepared to write as fast as possible. Bernie was sitting up in his bed and remained motionless as we conversed. Only the lights from the machines around him illuminated the area.

"Okay Tim, tell me where you're at"

"With?"

"The process. Your mom says you have to go through a medical board? When?"

"Six weeks, I think. The paperwork said they may change the date, but the initial date is in six weeks"

"Okay, that's not much time. Do you have a copy of your entire medical record?"

"The PEB Chief told me the board would have that already"

"Always have a copy of your medical record, Tim. Always. If you do not already have a full copy, that should be the very first thing you do. Once you do that, you should go page by page and write down a list of everything that is wrong with you and the dates they are mentioned in your record"

"Everything?"

"Everything! If you get organized now, it'll be a lot easier for you in the long run. Trust me, I had to fight for decades because I wasn't as organized as I should have been, but I didn't have anyone to give me the information I'm going to give you"

"Got it. Bernie, other than the medical record copies, what else do I need?"

"Tim, you should speak with everyone you know and ask them to write up a statement on your behalf. Having supporting documentation is important when you appear in front of the medical board. If you show up with just yourself, when they evaluate you, which they are doing the moment you step foot into that room, if you come empty handed, they will look at you as seemingly better overall."

"But I'm busted up still, they should see that, Bernie"

"Oh Timothy...did your mother tell you about the medical board I experienced?"

"Not really, Bernie. She just said you have experience with it."

"Do I ever! Timothy, let me tell you what I brought to my medical review board and what I would have done differently"

"Please do, I'm writing it all down, Bernie! All of it!"

Bernie shifted in his bed and straightened his blanket before he began. It looked as if he had to mentally prepare himself to relive it. I had my pen at the ready and began to write furiously as he proceeded to give spell things out.

"First, Timothy, I'm..."

Bernie stopped talking. I looked up at him and noticed he was looking towards the door, so I turned my head in time to see the door slowly close. There was enough light in the room for me to notice that Bernie seemed a bit uncomfortable at whoever was standing in his doorway. He slowly turned towards me and spoke at a lower, more hushed voice.

"Timothy, I don't want you think that I'm old and senile when I say certain things. I'm going to describe some things, sort of in code. Do you know any code, son?"

"Code? Me? No, Bernie, I don't know any code."

"Not actual code, son. Listen, when I describe certain things, I want to minimize the chances of someone listening in and making my life and your mom's life harder here at the hospital."

"Really? Who would…?"

"That's not important, Timothy, what's important is that you understand what I am saying by how I say it and the words I use. So, when I say something like: when you get out of the Navy, you will then have to deal with the Kingdom and the King is not a nice man. If I said that to you, do you understand what I mean by the term 'Kingdom'?"

"Would I be correct if I said we are both currently sitting within the Kingdom?"

"You would be correct, Timothy. Good lad!"

"Okay, I have the Kingdom down. I'll keep that in mind as we move on."

"Great. My board was in D.C., Timothy, but before I tell you about the D.C. experience, I want to tell you about a meeting that I attended, with other special warfare personnel, on base and private. It was put on by a gentleman who worked as a VSO Agent and put on this informative brief just for the special warfare personnel who were getting out. I am grateful to this day for that man, because he gave me and the others, insight into what to expect and how to avoid getting screwed over."

"VSO? What's the VSO?"

"Veteran Service Organization, Timothy. There are many out there."

"Which VSO was it, Bernie?"

"That's not important and I'll tell you why a little later Tim, let me keep going with this"

"Sorry Bernie"

"Don't apologize Tim, I just need to keep on track so you get the big picture prior to going through it yourself. Moving on...in this meeting, valuable information was provided to us, information that we would have never thought about, but made perfect sense as he broke it down."

Bernie leans up. "Are you ready to write this down, Timothy?"

"I am, Bernie, I'm ready"

"This gentleman looked at the dozen or so of us sitting in the auditorium and laid everything out, meat and potatoes, and didn't pull any punches. He said: You guys are tough guys, but what you don't realize is this game that's about to be played on you, is one that you are not prepared for. This game consists of people deliberately trying to befriend you and act as if they have your best interest in hand, but the reality is they're trying to lure you into a false sense of comfort so you'll slip up. Then they can write the report that will ultimately screw you over."

"Bernie, what was his advice?"

"I'm getting to that, Timothy, be patient. He continued: Let me tell you how it goes, you see, when you go to a medical appointment that the Kingdom sends you to for your disability exams, it will begin when you pull into their parking lot. They have you on camera. The appointment starts when you pull up! You get out and walk to the office. They check you in and you sit. Someone calls your name and you get up. Now, the person calling you in will deliberately walk fast to the exam room, because the average person will just follow and keep up. Their report will dictate that you kept up to their speed."

"Did that happen to you, Bernie?"

"Many times. He also told us that inside the exam room, the doctor or nurse or whoever is there giving you an exam, will tell you to bend over as far as you can go. He told us that us special warfare guys will bend over until we pass out."

"But that's what they said to do"

"No, Timothy, they just tell you to bend over as far as you can go. That was where we would screw up. He told us we would push through the pain to go as far as possible, not stopping when the pain begins, and that gives the person doing the exam the ability to write the assessment that you were able to bend all of the way over, or whatever degrees, regardless if you passed out from the pain."

"What? Did you experience that and if you did, what did you do to counter it?"

"It was because of that brief and his advice, which was to stop as soon as the pain begins or ask for clarification before you begin. So, when I was asked to bend over, I would first say "do you want me to stop when the pain begins" and they would tell me again to "just bend over as far as I could go" and I would repeat my question. They would reply with the same, generic and non-specific reply, so I would barely move and stop, and tell them that was as far as I could go"

"What would they say?"

"Each time the examiner would ask if that was as far as I could go. I would always say the same thing: "I stop when the pain begins, if I go any further, I could pass out from the pain". I would barely move because I was always in pain and it would become greater if I went any further"

"What did they do?"

"They would have to write down the degree of movement. The same goes for turning. I cannot turn my head much to the sides and they wouldn't specify for me to stop when the pain began. I would ask them the same thing about when they wanted me to stop, or if they wanted me to stop when the pain began, but every single time they would answer with the same thing: "just go as far as you can".

"Bernie, isn't it your word against theirs?"

"Timothy, my boy, technology today allows one to purchase a small, digital recorder that can record everything, and that, my son, is better than gold as far as advice goes!"

"So that guy warned you about these appointments but what if the person that calls you in from the lobby disappears?"

"When they would call my name, I would slowly get up because I'm in pain all day, every day and get stiff when I sit for any amount of time. I would hobble towards the person and when they took off, I would yell: "Pop a flare when you get there, I'll follow the smoke". They would usually stop and wait for me to catch up."

"Smart"

"I had my moments"

"What else did that guy tell you?"

"Have multiple copies of your medical records and associated paperwork. During this time the military was trying to save a dollar by getting people out when they could and I fit right into their time-frame. The main take-away from his brief was this: Trust no one. Your appointment begins as soon as you pull up into their parking lot. They will try to act like your friend to get you to do something that they could use to write their reports the way the Kingdom wants to see them. They will deliberately walk fast to get you to keep up, so walk at your own pace and ask for clarification on what they physically want you to perform so you know exactly what they are wanting from you."

"And have a digital recorder!"

"Exactly Timothy, have a recording device! I'm proud of you, you actually listened"

"Bernie, you are a treasure trove of information and if I didn't have this information, there's no chance I'd be as prepared as I should be. I can't thank you enough. Seriously!"

"Timothy, you're a good kid and I don't want to see you go through what I had to. My case wasn't even considered bad, I mean, I remember other veterans who battled against the Kingdom for 30 years and still died fighting them...I was one of the lucky ones and only had to battle for 20 years"

"Still, Bernie, 20 years? I mean...there's no way I want to have to battle for 20 years or even 10 years...seriously!"

"That's why you are taking notes, Timothy. Don't forget to write down statements from your friends, family and coworkers. The Kingdom has statement forms online that you can download and print a bunch of, to make it easier for your coworkers and friends to fill out."

"Got it. What do they need to say?"

"When I asked my teammates to write statements, the advice that was given to me was this: Make sure they explain what they have witnessed, difficulties with specific disabilities and how it impacted your daily life. The statements would vary with disabilities because I had so many, but each teammate wrote a statement explaining what they had witnessed with specific disabilities and my difficulties. I made copies of those also and brought them to my medical board."

"Statements. I'll have to jump on that when I get back"

"Make sure they put their email, their address and phone number, and at the bottom of their statement it has to state that it's a true statement to best of their recollection and ability. That's important. Do you have all that?"

"I do. Onto the med board if you're up for it?"

"I am. Are you ready, Timothy?"

"I am, ready to go."

"Okay, so what I did was this: I made sure I provided a full copy of my records."

"Even though they were going to have your medical records there?"

"Right, I brought a full set, this way there could never be a situation where they somehow had a partial copy and were unaware of it. I also brought a copy of the statements from my friends, family and teammates. I also brought the pain log that I created after my second accident."

"I don't know what that is, Bernie. What's a pain log?"

"Well, my pain log wasn't super-technical. What I wanted to track was how I felt in the morning and evening, each day, and note where the pain was and note the weather also."

"What's the weather have to…?"

"You will find out, Timothy. You will feel some aches that grow and grow and suddenly you will see the rain, or the snow, or the storm, but you will definitely feel it before it shows up. You'll get so good at knowing the weather, your friends will ask you rather than going online or watching the news."

"That's right, so how did you set it up? Did you use your computer? I know you don't like the computer, from what my mom had told me"

"I had a desktop computer that I did everything on. I did use excel but not like you kids can do today. I barely used it for what it could do, but I only needed it to track my pain on scale from 1 to 10. I would note what the weather was or what I did for a workout so that it might lead to a pattern I could track. If I could track a pattern, I might have been able to change things to reduce the pain."

"Did it work, Bernie?"

"It served its purpose. I tried to make it easier on anyone who needed to see it; like future doctors for appointments or whatever, by doing an average each week and each month. Anyway, I provided that to the board as well. I brought a lot of paperwork. I stacked my rolling case with the paperwork, had to be 30 pounds of papers, and dragged that to my lawyer. You should have a lawyer assigned for your medical review board."

"I think they assigned me one"

"Get to their office early. I got to my lawyer in the later afternoon, because that was the allotted timeframe, and I left her office at 7 p.m. My medical review board was to start at 7 a.m. the next morning"

"Wait, so you left the papers with the lawyer at 7 pm and your board was at 7 a.m. the following morning?"

"Correct. I left pounds of paperwork with my lawyer, who I know had it in her possession until 7 o'clock at night, because that's when I left. Each judge is supposed to go through the entire proposed paperwork prior to the board. This means each judge would have had to thumb through about thirty pounds of paperwork before the board began at 7 a.m. I'm pretty confident they did not do that. My board had three judges, but I'm not sure if that number changes depending on the board or if anything has changed since I departed the military."

"I'm not sur, Bernie?"

"Well, when I stepped into that boardroom, at 7 a.m. sharp, I was faced with three board members, who faced me from behind a long table that contained all of the paperwork I brought in the night before. The paperwork went from one end of the tables to the other. That was a lot of paperwork."

"Who were the judges?"

"A commander, a major from the marines and a female commander. The way the room was set up, I sat at the head of the table, which was the very end of 2 or three tables, vertically placed and they sat facing me, behind tables horizontally placed. I was a good 15 feet away at least. The commander asked most of the questions, the major took notes and female commander looked up once from her laptop throughout the entire board. My lawyer said her things, I answered questions honestly and the female commander showed how much she disliked me."

"Why didn't she like you?"

"Well, I couldn't tell if she didn't like me because I was a male, a white male, or a disabled white male from the Navy...but it was something within those parameters. She questioned my pain and was very crass to me. I kept my composure because emotion is not going to help you out in the med board. I sat, slight leaning forward as I usually do, because my back hurts, and when they asked me to stand and bend, I did as they asked. I barely moved. They noted it, the board was over and I left."

"What was the results"

"Well Timothy, this is where it got ugly. They came back with a 20% rating for my back and recommended a medical discharge. My disabilities then were extensive, but they were stating they were providing the rating just for my back and neck, and they deemed that 20%."

"That's ridiculous, Bernie. What did you do when you saw that?"

"First, I had to calm down, because that was a big slap in the face. I filed the appropriate paperwork and countered the statements that was in their report. In my statement, I added that by dropping off the evidence I provided at 7 p.m., there was no feasible way all of the board judges could have reviewed it all. I ended the submission by explaining the obvious bias from the female commander and her lack of attention to my degree of angle in the chair, since her comment was pertaining to my flexibility. The end result was my paperwork got denied. I submitted my final, last chance attempt, and that got denied. I tried, but my efforts went without. They held to their 20% and were not going to bend."

"That sucks. I'm sorry Bernie."

"Me too, Timothy, me too. I did some legwork after the fact and realized that 20% was an important number. If they rated me at 30%, they would have either put me on temporary retirement or full retirement, but at 20% they could medically discharge me and I would get nothing. This is why they chose the 20%...and I say that because I'm 100% and my back was the reason for the medical review board. Corruption all around. They needed more service members out of the military, they did what they needed to make it happen. Nice right?"

"That sucks. Makes me worry about my medical board. They didn't consider retiring you at all?"

"They did not."

"So, 30% they could have retired you completely, Bernie?"

"Timothy, at 30% they could have either retired me on the spot, or placed me on the temporary retired list...I think it was called TRDL for Temporary Retired Duty List, or something like that."

"What does that do for you, if they were to put you on the temporary list?"

"The temporary retired list, as I was told, meant I would have to get a physical every 18 months for a handful of years...I think up to 5 years, and if my doctor determined that I would not get better, then, at that point, I could be put on the permanent retired list. But if the doctor determined right away that I was never going to get better because my disabilities are a lifetime gift, I could be permanently retired immediately. So, it depends on what they decide."

"They gave you 20% and kicked you out. After all those years? I don't get it…when I joined, the recruiter hyped the military so much. I was bombarded with speeches about comradery and teamwork and how the military has my back once I'm in…only to find out how bad it is if they decide they want to reduce the force by a certain percentage!"

"Sorry Timothy, I don't mean to be the bearer of bad news or anything."

"No, it's okay, Bernie, I'm sorry you had to fight for your disability as long as you did. That really sucks!"

I had to stand and stretch and I could tell that Bernie was getting tired, so I closed my notebook, dropped it into my bag and turned towards the bed to say my farewells.

"Thank you again, Bernie, for everything. I have a lot of notes and will go over them tonight. If I have any questions, I'll write them down so I can ask you tomorrow…if it's okay that I come back tomorrow to get more information?"

"Come as early as you like, Timothy. I wake at 4 a.m. and most are still fast asleep for many more hours. Feel free to come by any time after that."

"Great, thanks again. Have a great night, Bernie!"

"You too, son. See you tomorrow."

I slung my backpack over my shoulders and proceeded to hobble across the room and out the door as Bernie tried to stifle a cough. I heard him, but I didn't want him to think I heard him. He sounded rough. I could hear the fluid in his lungs and the pain in his cough. I felt bad for him but knowing him the little bit I do at this point, I figure he wouldn't want me to make a fuss or call him out, so I didn't. My goal was to be prepared for tomorrow, as I was confident that Bernie was going to have me writing up a storm with more information.

I am not going to lie, I was quite depressed…well, depressed isn't exactly the right word to use…saddened? Maybe saddened is more appropriate. I was saddened that entire evening, thinking to myself that my case could end like Bernie's case had. It was almost a helplessness, because it felt like I had to battle an invisible army…which had stronger and much mightier weaponry on their side. I don't believe I slept at all that night and not because I was busy going over my notes. Most of the night awake was spent pondering what my options where if I did end up with the same fate as Bernie.

It was a long, long night.

3

TO BELIEVE OR NOT TO BELIEVE

CHAPTER 3
TO BELIEVE OR NOT TO BELIEVE

It was a rough evening, as I sat on my childhood bed and went line by line through my notes, I decided to rewrite half of the notes for a couple of reasons: first, it was for repetition, which would allow me to absorb it better, and secondly, my handwriting had to be deciphered due to writing too quickly. I wasn't going to arrive at the hospital at 4 a.m. because I was still on convalescent leave and was trying to get some sleep...which didn't work, but I tried. I rode to the hospital with my mother and arrived just after 6 a.m. I was more excited than the day prior and ready for another session of note-taking.

I stood by the nurse's station with my mom as she said her 'good mornings' to the other nursing staff, until she ushered me quickly to their break room to put her lunch away. I hobbled in and immediately took a seat to stay out of the way of the other, incoming nurses. My mom seemed bothered about something.

"Something wrong, mom?"

After letting out a sigh, my mom, Samantha, turns and sits at the break table across from me. She had tears in her eyes, which concerned me because my mom was really skilled at not showing emotion, perhaps because she was a nurse and needed to appear in control of the situations that presented themselves, or because she didn't want to concern me.

"Mom?"

"Timmy...I spent all last night going back and forth..."

"About what? Something wrong with me being here?"

"No, not really. I want to ask you something, but it's unfair of me to ask"

"What is it?"

"Well..."

"Is it about Bernie?"

My mom nods slowly and opened her mouth to answer me, but ceased immediately as two other nurses came into the breakroom to drop off their lunches. After a short greeting, both dropped their lunches into the refrigerator and left. I think they noticed we stopped talking as they entered and figured I was in trouble...at least that's what I would have thought. The door shut and my mom leaned forward.

"Timmy, I'm going to ask you to do something for me while you're here."

"Okay. Anything mom, do I need to bury a body or something?"

"Funny, Tim, I need you to keep an eye on Bernie for me"

"Keep an eye on him, why? What's going on?"

"I found some bruises on him yesterday...he didn't have those bruises the day prior"

"Where? Where were the bruises?"

"His arms and the side of his neck"

"Did you ask Bernie about them?"

"I did and he quickly dismissed them, as he commonly does"

"Does he bruise easily?"

"Not that easy...and not in different places."

"Is it possible he stumbles a lot when he walks around or something?"

"It's possible, but did he do much walking around when you were with him yesterday?"

"No. He didn't move much at all."

"I don't like asking you to do this, Timmy"

"Mom, it's okay. I'll keep a look out and if I see a bruise, I'll see if he'll talk with me about it."

"Thank you, Tim, I appreciate it."

"Mom...do you think someone is giving him the bruises?"

My mom gave me a sad smile, stood up quickly, stepped to the side and pushed her chair underneath the table slowly. She leaned down and gave me a gentle kiss on the top of my head, just like when I was a little boy. She held my head gently to her stomach for a moment before she exited the breakroom. I had to sit for a moment to let that all sink in, but I knew I needed to put my worries aside to see if I could spot any bruises on Bernie. Eventually I got up and made my way to his cave of a room.

"Good morning, Bernie! You ready for another fun-filled day with a kid who talks too much?"

"Of course, young man! I welcome the challenge!"

"You doing okay this morning, Bernie?"

"Good as can be expected, did you have a chance to go over your notes last night?"

"I did."

"Questions?"

"Not really. I had to decipher my handwriting because I was writing too fast at times."

"Need me to slow down?"

"That's okay, Bernie. I appreciate it. If I need to slow down, I'll let you know"

Apparently, I was fidgeting with my pen or avoiding eye contact, or something, because Bernie seemed to notice I was not acting normal.

"How are you doing this morning, Timothy? You seem preoccupied with something"

"Me? No, I'm good, just not happy about what I'm facing with the review board and all"

"Are you expecting the Navy to boot you out, as they did me? I wouldn't worry too much about that scenario happening to you exactly as it happened to me. I was a special case and it was a different time back then. Do you think the Navy is going to discharge you?"

"I do. I saw the x-rays and my mornings are rough. I know the pain should subside, but I don't think my life is going to be pain free...Uh, Bernie?"

"What's up, young man?"

"I've heard from so many people at the hospitals and clinics that I shouldn't worry too much about things, the V...uh...the Kingdom will take care of me when I get out. It feels as if they're trying to get me to just focus on leaving the Navy and not worry about them or the Kingdom screwing me over on my way out."

"Well, the Kingdom won't screw you initially, and by that, I mean: you will have to wait some time for the King to process your claim and rate your disabilities. This is why it's important to be very well organized."

"I get it... I'm just worried that these people are trying to get me to not ask questions and just trust their word, which I don't."

"That's good, watch out for number 1, Timothy. You have to cover your own back during this process. No one else is going to ensure they don't screw with you, or deliberately mishandle your paperwork…I have horror stories on that subject also…but let's go over what the process is if the Navy were to discharge you, okay?"

"Sure, Bernie. Do you mind if I lift the shade a little to let some sun in…when it finally makes an appearance?"

Bernie points towards the corner, where a lone lamp resided.

"Sure, or you could drag that lamp over to your seat and use that. Or both. It doesn't matter to me. I just can't have the blinds fully raised. It makes my head hurt because of my pupils."

"What's wrong with your pupils?"

"Nerve damage. One pupil is larger than the other and it lets more light in. That light feels like a laser drilling into my skull. It's not a pleasant feeling."

"I bet. I'll drag the lamp over."

I grabbed one crutch and hopped over to the lamp in the corner. I unplugged it and used it to help me cross the room easier. I set it down next to the recliner and plugged it in. I turned the lamp on and stood on one foot for a moment while watching the bulb brighten. Convinced it wasn't going to get any brighter, I turned to Bernie.

"Is that okay, Bernie? Too bright for you?"

"No, that's okay the way it is."

I noticed Bernie's eye was swollen a little and his cheek was red on the same side.

"Do you have allergies, Bernie?"

"No, why?"

"No reason, just curious. Your eyes are a bit puffy. I have allergies and sneeze a lot during allergy season. If you want we can ring my mom and have her bring you something"

"I'm fine, Timothy, thank you for your concern. I just slept wrong."

"Okay"

"Where do you want to begin today, Timothy?"

"I guess from getting the letter that says I'm going to be discharged. That's probably the best starting point"

"I don't know what they call the class you take when you are preparing to depart the military...in my day it was called a TAP class. Transitioning something something. It was a class that helped you work on resume writing, job fairs, interview techniques, a dress up day and then a day where you can speak with a VSO."

"The VSO...Vet...?"

"Veteran Service Organization...the groups that exist to help veterans with their disability claims. There's a few of them...American Legion, Veterans of Foreign Wars, AMVETS and some others. Now I think there are state driven VSOs for the veterans in their states, so you have options."

"The VSOs are at this class with us?"

"At the end, or they may do it differently today, but for the TAP class I attended, the VSOs were there at the end of the class...and it was a mixed bag of opinions from some of the other service members who were in the class with me."

"Which ones were at your class, or were all of them there?"

"I don't remember...but I remember the experience. I brought my medical records, as instructed and we each had a time slot to meet a VSO. I went in during my slot and sat down. The guy behind the desk reacted like it was a pain for him to be there. He grabbed my first record, commented that I had a lot of documents, then flittered through the pages like he was shuffling cards. He would stop at a spot, scribble something down on his notepad and continue flittering through the pages. After about 5 minutes, he pushed my records towards me and handed me the paper he was writing on."

"What did the paper say?"

"He listed 5 items. Just 5. Two volumes of records and he listed 5. He told me those 5 things are what I need to claim because nothing else in my records are worth claiming."

"What did you do?"

"I took my records, stood up and told him I apologize for wasting his time and I left."

"That's awesome!"

"Not really, I could've gotten in trouble, but thankfully I had the information I needed to sort of navigate through this mine field. You have to choose a VSO to represent you. The good thing is there are some VSO agents out there that truly care about helping veterans and work really hard for the veterans and their appeals. Unfortunately, those great VSO agents get inundated with veterans who want to be treated better than they have. Word gets around fast."

"Are they all like that, Bernie?"

"I don't think so. Remember the guy I talked about yesterday who came by and briefed us in a closed session?"

"I do"

"Well, he was an agent, which is why I went with that VSO...unfortunately he was forced to retire because they were downsizing."

"That's horrible. Man, Bernie, you can't catch a break!"

"That's what I said, but I had a final meeting with him and he gave me valuable advice. He told me that I needed to make sure I kept a copy of everything the VSO does, and to request a copy of whatever they sent the Kingdom. This was to ensure my records were up to date and accurate. He left and warned me about how veterans flood to those agents that do the most for the veterans."

"That makes sense."

"It does, which makes it hard, regardless who you choose to go with. I kept with this organization until I finally had enough. I received yet another rejection from the Kingdom and was assigned a different woman, because my agent wasn't present for whatever reason. Anyway, I went into this office and this large woman sat behind the desk and glared at me as I entered. I said good morning and she commanded me to sit. No good morning, no hello, just sit."

"That's rude"

"It gets better...she quickly shoves her notes aside and puts her hand out. I look at her hand as I sit in the chair and apparently, I looked pretty confused because she barked at me to give her whatever paperwork I needed her to see. I gave her the envelope with the denials and told her I had some items of concern. She looked at me quickly and said: "I have time for 3.""

"What? 3 what?"

"Three concerns. I replied that I actually had 5 concerns and they were all equally important. She wasn't having it...and at that point I was so frustrated from her treating me like garbage and the Kingdom runaround, I stood back up, grabbed my envelope out of her hand and I apologized for wasting her time. I turned around and I left. I didn't look back, that was the last time I stepped foot into that office."

"Is that your only experience with VSOs?"

"No, but each story seems to end the same each time. Eventually I took my own case in hand and did my own appeals."

"So I can..."

"Don't be your own advocate, Timothy! Don't do it! I believe the saying is: A person who represents himself has a fool for a lawyer!"

"I've heard that before. But how do I know if I have a good one? A good VSO?"

"There's no way to tell, Timothy. You could ask around, go to hospitals inside the Kingdom and talk with veterans who are there for an appointment and see who they use. That's probably the best way to find out which agents are the best. You could visit the VSOs also and ask for references from the veterans they have helped. If they won't give you references, you may want to find someone else."

"Makes sense"

"The main point, Timothy, is to do your research before committing to a relationship with anything or anyone. It's like buying a car, you're not going to walk onto a lot, point at a vehicle and tell the salesman that you'll take it before you do your research on the vehicle. Right?"

"You're right, Bernie, it makes sense. They make it sound like once you become a civilian the whole world opens up for you, and by they, I mean everyone involved in the medical board and discharge side of things."

"That's a correct statement right there! Learn from my mistakes and do your research"

"So...what should I be looking for with the VSOs, when I do the research?"

"Timothy, I did a lot of tracking and digging on the VSOs that exist, along with non-profit organizations. The NPOs I was giving my money to, turned out to be scam artists and greedy people who value profit over cause. I had a big problem with that. Charity navigator online can get you started for NPOs. If you want to see how much money the VSOs and NPOs bring in, it's available online in tax returns. It'll wake you up to the greed that exists."

"Bernie, did you ever become a member at any VSO? I've seen posters up all over the base for some VSOs, but I don't know of anyone personally who belongs to any"

"I was an officer at one VSO, for a couple of years, but I was so aggravated at the complete and utter ignorance from the top and their lack of focus towards helping veterans, I pretty much stopped giving them my money and I definitely stopped making them more money."

"How do they make money, if they are not for profit?"

"They spend it on paper, Tim, they spend it on paper"

"So, they lie?"

"I'm not going to comment on whether they lie or not, because I want you to gain your own opinion from your research. It is up to you to believe or not believe what you see. That's why you do your diligence in research. If you don't, you'll spend years of your life battling back to the sunlight, like I had to do."

"Okay, so do my research. What if they convince me that I should sign with them as my VSO and they turn out to be junk? What do I do then? Can I fire them and pick another?"

"You won't be in the military any longer, Timothy. This means you don't have to take what someone has to say and not have a voice. You get a voice. You get to find another one, then fire the incompetent ones before you sign with a new VSO. Judge each by their efforts on your case"

"Are there warning signs, Bernie? Other than what you've told me already."

"Lots. If you look out for them."

"Such as?"

"Such as, well...if the VSO refuses to supply you with a copy of your profile in their cabinet. If they tell you something is not worth claiming without showing you that information on paper, through the manual or other legitimate correspondence. Telling you they don't have time for your concerns. Lots of warning signs exist, you just have to stay awake to see them."

"Got it"

"Do I have to pick one at TAP class?"

"I don't think so. If you know you're getting out, now is the time to start your search and speaking with representatives at the different VSOs. This way you can choose one before getting out. Or you could go with a law firm."

"Law firm...so sue? I don't understand."

"Well, once the law firms figured out they can make 30% off of each veteran's lump sum award, the lawyers scrambled to get their agent certification through the Kingdom. Then the law firm can represent you and do the fighting for you."

"But?"

"But they will snag 30% of your check"

"They get 30% of my disability check forever?"

"Not forever, but if your case goes on and on and the King finally rates your disabilities appropriately and you receive, say, $100,000 retro check, the law firm would get 30% of that $100k and you'd get the remaining."

"That's not horrible...I didn't have the money anyways, right?"

"That's what most veterans think and to be honest, I needed to use a law firm also, but I learned some things with the law firms as well. First, there are certain rules that you need to know about, such as if the law firm fires you, then you pay them for the hours they invested into your case."

"Why would they fire a veteran?"

"Well, the only reason I can figure is if the veteran was unwilling to let them do their jobs and was constantly harassing them. I cannot see a law firm firing a veteran just because they can't get any more money from them...but...that doesn't mean the law firm will continue to fight as hard as they should for your case once they get the big payoff"

"I don't get it"

"In my particular case, I signed with the law firm. They talked it up big, but I didn't really have a choice...I was my own advocate and you remember what I just said about being your own representative, right? Well, I hit a brick wall and couldn't get over it. I looked up the list of approved law firms and reached out. I was rejected by 4 of them because they wouldn't receive a big payday."

"That garbage, Bernie, why..."

"Timothy, it's okay, really, I am straight and up front with my conversations and I started each with the statement that it wasn't going to be a big payday for them. I then gave some details and expressed honesty in their approach and ended the contact by stating: if they didn't think it was a case for them, to let me know and I'll move on. I received 4 no's and moved on. You can't dwell on things, you have to move forward. Always move forward"

"Progress."

"Progress is right, because even small progress is still progress."

"But these lawyers ended up screwing you over, also?"

"Not really, they were smart about it. They fought and I was finally awarded a rating that they could capitalize on and make some money...but I had other paperwork awaiting decisions as well, and I reached out the firm to speak with my representative and was finally able to speak to him about that topic. I was told the law firm was a business and they still need to make money and my other issues were not as important as a newly discharged veteran who was only granted 10%."

"What?"

"That was the point where I informed the lawyer that I didn't sign a contract with that other veteran, I signed it with his law firm. I also stated, in very emphatic terms, that their management should send me the release paperwork if they are done fighting for my case.

I said a bunch more, very heated things, and figured that would be grounds for them to fire me from the contract, so I was prepared for that."

"Did they fire you?"

"They did not and I believe they didn't because the last appeal that was submitted by the firm included the transcripts from the medical appointment I had, concerning a denied rating I received from the Kingdom! I'm pretty sure that office had the proverbial lightbulb go off over their head, which was their aha moment of remembering that I record everything, hence the transcripts. I am pretty sure they realized that I must record all conversations and since they did tell me that they are a business and still have to make money AND that my case is now not that important and tossed onto the back burner, that I must have captured that on recording also."

"Oooh!"

"Yeah. It's just a guess, but I think it's a good guess that they figured I recorded that conversation. That wouldn't make them look good at all if that recording made it into the local or national news that another law firm had screwed over veterans."

"Good thing you record everything, Bernie. I will definitely purchase a good recorder and record everything from this point forward. It's obvious how important it is for everything."

"I cannot tell you how right you are, Timothy. You need to highlight the important parts of what we discussed because you definitely need to make sure you don't forget that information. It's important for your sanity and time. We don't get much time on this planet and wasting it on this subject is definitely not worth it. Have your ducks in a row and close it out while you're still young."

"I get it, Bernie, thanks!"

"The most important point to this subject, Timothy...the law firms won't take you on, as a case, until your claims have been denied."

"So I need to apply first and then get denied?"

"That's correct...but check with them first to make sure...it could have changed."

"Again, thanks Bernie."

"My pleasure, now if you'll excuse me, I'm pretty tired. I should probably get some sleep before my shows come on."

"Okay Bernie. Anything I can bring tomorrow? Is it okay if I come tomorrow, or have you told me everything?"

"Timothy, we're scratching the surface with this, barely. More tomorrow, but make sure you read over your notes tonight, for anything I need to expand on tomorrow. Lots of information came your way today, I don't expect you to absorb it all, but it'll make better sense if you go over it a couple of times to ensure you absorb as much of it as you can. Okay?"

"Got it Bernie. You sure you're doing okay? Is there anything I can get you or do you want me to get my mom?"

"Nope, just tired, Timothy. I just need some rest. I'll see you tomorrow. Have a good night, Timothy."

"You too Bernie, you too."

I could tell that Bernie was hurting and it seems he got really tired, really fast. I also noticed that he was favoring his arm and moving only a little. One thing with Bernie is when he got excited about something, he became animated and with the information he gave to me today, it seemed like he was in too much discomfort to be himself. It could've been me spending most of the time frantically scribbling notes as he spoke, but it was the same as the first day I visited with him. It was clear he wasn't feeling his best.

DISCHARGE WOES

CHAPTER 4
DISCHARGE WOES

It was a particularly painful morning. I could barely move out of bed because of the storms that were raging outside my window, but I forced myself up. I sat up, dangled my legs over the bed and listened to my mom hum her favorite song out in the kitchen, while she was cooking up our breakfast. I winced as I tried to stand, but the pain forced me back onto the bed, where I remained for a moment to contemplate how life was going to be for me in the future. I wondered how Bernie was able to survive for as long as he had, especially with his disabilities.

I stretched everything out while seated on the edge of my bed and heard my mom step up to my door, to wake me. I hollered out "I'm awake" as she began to rap on my door and I could tell she was smiling by her "okay smarty-pants" comment back. My mom was a worrier so I didn't want to look like I was experiencing more pain than normal. She was a nurse and very observant to all things health-related, so I needed to be a good actor, especially today.

I hobbled out to the dining room and took residence behind a heaping plate of scrambled eggs, bacon and toast. A big, hot cup of coffee soon accompanied the gourmet breakfast as my mom sat next to me to eat her breakfast. I could tell something was on her mind, because she always piles more food onto my plate than she knows I can actually eat. My plate was overflowing with food. Something was bothering her.

"Thanks mom, how're you this morning?"

"I'm fine, Timothy, how's your leg doing?"

"It's okay, just a little bit stiff."

"Mhmm, is it now?"

"Yup, but I stretched it a bit while I was sitting on the bed, so that helped."

I dared not look up because one look and she would've known for sure I was not telling the truth. All she needed to do was make eye contact, then she would have known for sure that I was in a lot more pain than I wanted her to know. She was good at telling if someone was telling an untruth, she had to deal with it constantly at work.

"Timothy?"

"Mom?"

"How was Bernie yesterday?"

I set down my fork and took a sip from my coffee, as I glanced at my mom. She looked at me with a look I was unfamiliar with, almost like she just realized I had grown up and actually drank coffee like a big boy or something. It was odd and unsettling. I set my mug down before answering, because I didn't want to alarm her.

"Mom, do you think something is up with Bernie?"

"Maybe. How was he?"

"He was okay...not as animated as before, but okay. He told me a lot of things. I was writing like a madman yesterday. But..."

"But what?"

"Well, it might be nothing."

"How about you tell me and I'll decide if it's nothing?"

"Well, it looked like he was favoring his arm and he barely moved. I don't know him as well as you, but comparing how he was on the first day I met him, it seemed as if he didn't feel as good or something."

My mom took a sip from her coffee and stared down at her plate of eggs. I hadn't notice until that point that she didn't touch any of her food. I felt like I needed to sate her concerns.

"Mom, Bernie and I are continuing our conversation this morning. If he wasn't feeling good, do you think he would still continue to welcome me to come back each day?"

"He would, Timothy. He's a great man who has always been selfless, no matter how he felt...that's what I like the most about him, but also aggravates me the most."

"Where do you think the bruises came from?"

"I don't know, but I have an idea."

"Where?"

"That's not a concern for you, young man. Your concern is to finish your breakfast, take a shower and get dressed so we aren't late for shift!"

"I don't have shift, mother, so I can't be late!"

"Okay smarty pants, my shift! Go! Get into the shower and get dressed, we leave in 30 minutes, whether you're ready or not! Come on, mister!"

I could tell she was still concerned about Bernie, but she played it off rather nicely, so I needed to let her have that moment. I know she was concerned about Bernie's health, but also my health and that's a lot to put on her plate. Not that I believe I fooled her, but I think she appreciated the fact that I put her mental wellbeing ahead of my physical and tried to make her worry less about Bernie. I put that thought into the back of my mind and got ready. I was sure it was going to be another big day of writing and information with Bernie, and I was not disappointed.

I arrived at the hospital that morning with the same excitement as previous days, but this particular day I brought a gift for Bernie, to show him how much I appreciated all his help. The evening prior, I went up to the attic and rummaged through some old boxes that contained my grandfather's military ribbons, pictures and awards. I found his pins and figured it would be appreciated if I brought Bernie something that meant something to me and my family.

I hobbled quietly down the hallway towards Bernie's room, as usual, and lightly rapped against the door as I stepped inside the darkened cavern. Bernie was slightly hunched over with his eyes closed, as he sat in an elevated position. I stopped suddenly, afraid that he was still sleeping and not wanting to wake him, but my crutch thumped against the floor loud enough to gain Bernie's attention. He glanced at me and nodded as he settled back into a somewhat reclined position, and forced a painful smile.

"Good morning, Bernie. Is it okay if I come in? You awake and all?"

"Yes sir, come in Tim. I was just stretching my back."

I hobbled my way over to the chair, as usual, set my bag next to the chair and sat down. I reached up and clicked on the light next to the chair, then withdrew the notebook and pens from my bag. I pulled out the old, cardboard box containing the gift I brought and set it onto my lap. I looked up at Bernie and his eyes were glassy, and I noticed that there was a dark ring around one of his nostrils.

"Bernie, are you doing okay this morning? Is your nose bleeding?"

Bernie wipes at his nose and looks at his sleeve, then he looked up at me and that was when I could see his bloodshot eyes. He looked like he hadn't slept at all. He forced a smile and nodded his head before lifting his sleeve towards me.

"Nope, nothing there. Air is dry in here, probably from last night while I was sleeping. What's in your hand, Mr. Timothy?"

I stood up and stepped over to his bed and set the old cardboard box onto his blanket. I turned and returned to my chair.

"Mr. Timothy, what is this?"

"It's a gift, Bernie, for helping me out. This means a great deal to me and I appreciate it a lot. I wanted you to have something that belonged to my grandfather."

"What is it?"

"Open it. It's nothing big."

Bernie winced as he brought his other arm up to help open the box. He brought it closer, opened it and retrieved the dull, pewter item. He grinned big and he lit up for a moment before setting the item back into the box.

"I can't accept this, Timothy"

"Why? I want you to have it, Bernie!"

"Your grandfather was a paratrooper? I can't accept his wings"

"Bernie, you were airborne and he was airborne. You deserve those wings! That's history there! I can't think of anyone more deserving than you, Bernie. Please. Please take those wings. They'd be sitting in a box until the end of time, collecting dust and living in the dark. This way you can make all the nurses go ga-ga over your new pin!"

Bernie smiled and retrieved the jump wings, removed the backs and mounted the wings onto his t-shirt. Then petted them gently.

"This is quite the gift, Timothy. Thank you."

"Thank YOU, Bernie! You're giving me valuable information that's going to save me a lifetime of aggravation, stress and pain.

I wanted to give you something that could express how much I appreciate all of this information. Truly."

"Well...you're welcome, but it wasn't necessary to get me anything. Anytime I could help someone avoid what I had to experience, I gladly go out of my way to help them out. So, thank you, Tim. I'll treasure this always and never take it off."

"Great, you're very welcome. I'm glad it's in a proper home and appreciated. So, what can we discuss today? Do you have any experience with college stuff or the vocational thing for veterans?"

"I have a little bit of information regarding the college topic through the Kingdom, and some stories to enhance the information, just so you understand what I went through! Mostly because I want you to look up the information to get the most current information on how they do it and what it entails."

"Awesome, let's do it."

"Later, actually, I think it's important that we discuss what happens when you leave the military."

"I don't know what you mean. Regarding what, Bernie?"

"Tim, when you get out, you've been used to things occurring at certain times, by certain people, and you had a role to play. Once you get out, that ends...not entirely, but it is going to be a bit of a culture shock. Has anyone talked to you about this?"

"I don't understand what 'this' is, Bernie. What are you talking about, specifically?"

"I'm talking about medical, treatments and how long you're going to have to wait until you get the King's acceptance and possible problems, like I experienced, at the hospitals within the Kingdom. It may sound dull, but it's very important to know so it doesn't come as a surprise. You won't be disappointed, especially after you get out. Trust me."

"I'm in your hands, Bernie. Talk away."

Bernie painfully adjusted in his bed and winced, but he tried to hide his discomfort by adjusting more. I acted like I didn't notice, but I did and was concerned even more so now that he was in trouble. He started in on his experience so I kept my head down and tried to keep up with my note taking.

"Okay Tim, you will need to check out of your command and they're going to send you around to the supply locations and command locations to gain signatures and turn gear in. After you do that, you'll have to go to your Personnel Support Depot, where your service record is kept. I suggest you go there before you depart the military and make a full copy of your service record! I'm not sure if you're going to be able to do that or not, so ask your administration office at your command if they'll request it for you, then make your copy of the entire record. With me so far?"

"I'm tracking, Bernie, go on!"

"I'm sure a discharge physical will be in order and it's very important that everything gets looked at and recorded properly. If it's not in your records, it won't matter to the King, when you file your claim. You want to make sure every problem area gets attention, and don't be macho and think that you don't feel that bad now!"

"I wouldn't..."

"Timothy! We all did and still do, but you don't! Got it?"

"Got it."

"Good! So, you need to make a full copy of your medical records, as I've mentioned before, and you need to comb through it, page by page, and list everything that is in your record that is relevant. That is what you're going to ask the physician to look at and note. The Kingdom is going to rate what they have proof of, otherwise you will be spending a lot more time trying to justify any, and all, disabilities."

Bernie held his hand up to me.

"Matter of fact, you need to know that whatever you don't put into your initial claim, will be really hard to claim and prove down the road...you need to make sure you put everything down on your initial claim, no matter what your VSO tells you, and if you rely on the VSO to put the list out, you make sure everything is there!"

"Will the VSO talk me out of them?"

"Perhaps, or they could make you feel like you wouldn't get rated for this or that, or your claim will take even longer because you put so much down...the point being...you put everything down for the initial claim, otherwise it will be hell for you to gain any ratings on something you remember after the fact. A lot, and I mean a lot, of veterans, when they get out, say to themselves that they feel fine at that moment in time, and they never do anything. Then, fast forward a decade and they are completely falling apart and in great pain and in need of surgeries. Do you think the King will make any effort to look in their records to see if it was something that was connected to service?"

"Um, no?"

"No is right, Timothy, and don't you forget it. Ever! Seriously, I've spoken and helped out many veterans who realize after the fact that they should have submitted everything in their initial claim, some of those veterans are still battling for coverage after 3 decades...30 years! You don't want to be that person, Timothy, trust me."

"I believe you!"

"Good. Now, some veterans like to think that if they file a claim and get disability AND they feel good at the moment, that they would be taking money out of the pockets of other veterans. This is something the KING has NEVER tried to correct for accuracy. This means the KING allows that lie to continue because it means more money in their pockets. Their bonuses get bigger and budget larger and larger! They have one of the largest federal budgets in the U.S."

"Something like $150 billion dollars, right?"

"At least, Timothy, at least. The way it works is this: you are locked in to be discharged on November 1ˢᵗ, let's say, so the Kingdom has a number of expected departures and they keep track of the trend of discharges of past years. They do this to estimate the amount needed to pitch the Congressional Committee in D.C. when they are in front of them trying to clear their fiscal allowance of funds. October first starts the new fiscal year, so the KING is in front of the committee well before that, to get the approval and the monies needed. I've watched a few hearings and screamed at my computer each time. It is aggravating!"

"Because of the amount of money they're asking for, Bernie?"

"Well, not just that but also because the Committee will ask them important questions and if the KING himself doesn't want it blasted live on a public venue, the King will tell the Committee members that an answer will be sent to their office later, or claim complete ignorance and uses an excuse of needing to check in with someone to get the specifics. They would then schedule another meeting with that committee member to discuss the specifics. This means zero answers for those paying attention and watching these committee hearings. I've seen it happen over and over again, and it's aggravating because these politicians and the King himself, believe we are idiots and believe everything we hear and see without researching ourselves."

"Holy crap!"

"Indeed, Timothy...I could go on and on about tracking the Congressional Committee bills in the House and Senate, but perhaps another time. That was something I did constantly, to watch how these politicians who state how much they love veterans, turn around and propose dozens of bills only to let them die away, but they don't tell you that's why they proposed them, they just tell you how much they did for the veteran community. It's aggravating to say the least."

"I bet. I don't think I could do that, Bernie."

"Nor should you want to, but it's something I did because I needed to be sure of my suspicions, and I wasn't disappointed. I tracked and maintained the proof that I was right. That was aggravating also. Anyway, enough of that, let's keep going. Where was I?"

"Getting out..."

"Right. Okay, so you make sure everything is claimed from your record and submitted. When you get out, no one tells you that you're still covered for a period of time. When I was discharged, I wasn't told anything, I handed in my ID and told to leave. That was that. I had my copies of my medical record and I ended up getting bit by an insect and needed to go see a doctor. I brought my medical records with me, expecting the doctor to want to see them, but was wrong about that."

"What'd the doctor tell you?"

"He picked up one volume, thumbed through it quickly and tossed it back onto the tabletop as he sneered at me while stating how much he hated seeing military medical records. I had to tell him that was all I had and he informed me that he's never seen a complete record that would satisfy civilian medical professionals. I had to just shrug and ask him what was next. Well...before that I had an interaction with one of the veteran's hospitals...sorry, let me back up a bit..."

"It's okay, Bernie, tell me everything!"

"I was discharged, had to submit my ID card and left the base. I tried to go to the closest veteran hospital and register, so I could get my primary care doctor, but that meeting was far from pleasant. I was told they couldn't see me until my claim was processed, which would be about 3 years. I wasn't having that. I quickly informed them that I was medically discharged and awaiting the claim results and I needed pain medication and the Kingdom was supposed to take care of me. Remember how I told you that you'll be made to feel like the Kingdom will magically care for you after you depart the military?"

"Yeah"

"It's a lie! I was quite angry at this guy and expressed my discontent. He told me to contact the King, which is a clear sign that I was going to get the run around. So, while I waited, I went to this doctor to ask him to be my primary care physician so I could get my medications, until the Kingdom finally takes me in. This doctor was going to have to put me through an entire workup in order to take care of me, which I was okay with, but the timing was going to be bad for my chronic pain. I reached out to my doctor on base, at my old Team and this guy was an incredible human being and always took care of me. It was no different after I was discharged and that was when I was told I was covered for 6 months after I was discharged."

"6 Months?"

"You need to verify that, Timothy! Ask someone in your admin office if you are covered by the military medical system for a certain timeframe after you get out. Then ask what that timeframe is and what services are available. Find out everything you can about it, to avoid everything I had to battle."

"Got it!"

"It was a jarring and aggravating experience for me that I shouldn't have had to go through, so hopefully this will save you from experiencing that as well. It is a culture shock, Timothy, to get out and to be around civilians who do not have the same mindset about being on time, or having a strong worth ethic or sacrifice."

"I'm not looking forward to getting out, Bernie. I did work with some civilians on the fire department on base and they were okay."

"You're going to meet all kinds, I promise you this. So, my best advice to you on this subject is: don't expect everyone you meet to be like everyone you have ever met. That will save you from great aggravation."

"So, lower my expectations? Got it."

"Good lad. You nailed it. You should be realistic, though, and realize that no matter where you go, or what you do, you're going to have good people who work really hard and barely get recognition, and people who hardly work and seem to get all of the praise and benefits off the hard work of the good workers. You can't avoid it, unfortunately, but you can embrace it and try to be the best you, and not worry about everyone else. Just maintain awareness of what others are doing around you. That'll help keep you safe."

"Got it. I'm a pretty hard worker, I think."

"Your mother seems to think so, Timothy."

"Well, she's my mother and is supposed to say that"

"Your mother is a very strong woman who speaks her mind and I'm sure if you were a slacker, she would've told me when she spoke about you. She gushed about you each and every chance she got."

"Yeah, sorry."

"No need to apologize son, it's a good thing that she has so much faith in you and is proud of what you do. I never had that and am jealous. Be proud and happy you have such a strong support foundation in your mother, she's going to be your greatest strength."

"That's true. She is great and she's a hard worker also. I think that's where I got it from."

"Well, how well you do is a reflection of how well she did raising you and I have to say, she did a great job there. You seem to have a great head on your shoulders and determined focus, which most do not have today. It's refreshing to see something like that and it'll help get far in whatever it is you want to do with your life."

"I thought it was going to be the military...but that obviously wasn't going to work."

"So, it wasn't in the cards, so what? Move on and don't dwell on what you cannot control. That'll serve you well to absorb right now."

"What?"

"To not dwell on what you can't control, Timothy. There's nothing you can do about your upcoming discharge, other than ensure you don't get screwed over by listening to this old man's stories about getting screwed over."

"That's my hope...to not get screwed over."

"That's a good plan, but don't dwell on things. You're going to look back and have moments of sadness that your past goals were cut short and not because of something you did, but something that happened to you. You had no control over the roof collapsing on you, you couldn't control that. It was in the cards, Timothy, and you need to embrace that it happened...but it happened in the past, and that you cannot change. Do you understand that? You don't have to like it, but you do have to embrace that fact because if you don't, it could very well eat you up inside."

"I get it, Bernie."

"Do you, or are you just saying that because I'm putting you on the spot right now?"

"No, I get it. I do. I don't like it, as you said I don't have to like it, but I do need to embrace it as a fact and move on with my life. I get that it will haunt me, probably forever, but I still cannot change what happened."

"Exactly. It's okay to feel sad or disappointed. It's normal, actually, to feel that way, otherwise it wasn't truly a dream of yours."

"I don't understand."

"Timothy, if you went into the military just to do something, anything, because you were bored with life, and then the accident happened, you wouldn't be disappointed because you didn't set your sights on the future and work hard to get there. Instead, you wanted to be a firefighter, you wanted to retire from the military and make it your world. This is why you feel disappointed and sad at times. That's okay and that, my good sir, is what normal looks like."

"My normal, which is probably not like everyone else's normal."

"It shouldn't be everyone else's normal. Your normal is your normal, son. None of us are actually normal, but for us...it IS our normal. The goal is to find others that you enjoy their company, who are your type of normal. Match your normal and you get good friendships and relationships out of it...that lasts a lifetime."

"Bernie, do you still have your type of normal friends around?"

"Do you mean are any of them still alive?"

"No, I didn't mean..."

"No, it's a valid question, Timothy. I was messing with you. Most of my brothers are long dead. Some died doing a hero's job, some died because of cancer and others just disappeared, which is what most of us wanted to do anyway."

"Why? Sorry, you don't have to answer that, I don't mean to pry."

"No, it's fine Timothy. It's important to gather information. That's how you'll know who is who and what is what. Most of my mates are not mentally stable and it's partly because of the injuries we incurred while doing our job, but regardless of the injuries, we loved the job and the comradery. We had idiots and douchebags that worked with us, you can't avoid that. The hope is that you bond with select individuals who you know will have your back no matter what. I'm sure you had some buddies at your firehouse that you could trust with your life."

"Well, we're supposed to trust them all."

"But did you, Timothy? Did you trust them all with your life?"

"No sir, I can't say I could"

"Now you understand? But let me ask you this: were there some within the firehouse that you trusted enough to tell them things you haven't told others in your life? Not even your mother?"

"There were."

"Then you were a rich man, Timothy. Some never get to have that in their lives."

"They were the ones that actually went back inside and pulled me out."

"Was the fire out at that time?"

"No. They were actually by the truck monitoring the gauges when they saw the roof collapse. They ran inside specifically to get everyone out."

"Were there others trapped?"

"Bernie, I don't even know how they found me. I was last man on the hose that fire, which is unusual because I was usually closer to the nozzle, but we had a man out sick so I needed to change positions. The others were further ahead and were able to escape safely."

"None of them came back inside? To find you?"

"No. They scrambled to get out of the building. When they stumbled down the alleyway, they passed my buddies who asked each one if they had seen me and where was I on their hose team, and each person told them that they didn't know. A couple said they thought I had already escaped, but my buddies knew me and knew I wouldn't have abandoned the hose team."

"So, they came in and found you, rescued you?"

"Yes sir."

"Please stop that sir crap, Timothy, I worked for a living!"

"Sorry, right. Yes, they found me trapped under the roof portions and dragged me out. I was unconscious."

"Did they visit you in the hospital?"

"They took turns sleeping in my room. They were the only ones that did that. I got flowers from the station, and some visits, but for the most part I only had two buddies that actually showed concern for me and my situation."

"Timothy, those friends will be with you forever. That's exactly what I'm talking about. Good, I'm glad we had this conversation, Timothy, it's pretty important because what you consider normal, is not what others will consider normal. You'll see people lose their minds because they have to wait 20 minutes in line, whereas you could stand waiting for hours and think that's normal...because that's what you experienced in the military."

"It sounds like a lot of work...to accept civilians and all, Bernie"

"It is work, but eventually you'll acclimate to it, but you need to remember that at some of these locations you visit, it could have veterans working there. This doesn't mean these are nice people, or people who take pride in their job, or even people who care about other veterans. Don't confuse someone else being a veteran as some-one who actually has core values."

"I never thought of that. That makes sense."

"Unfortunately, it's something that will aggravate you also. Same thing goes with the medical professionals that you'll have to visit for disability claims. Some of these doctors will make you question a lot of things."

"Any chance you could give me some examples on that?"

"How about after lunch, Timothy? I feel like you have a grasp on the importance I put on what to expect when you get out. I can only give you my experience in the hopes that it'll give you a heads up on what to look for, in case it comes at you."

"I think I do, Bernie. Is it possible to get an example, one that I could relate to?"

"Okay...here's an example that I think you can appreciate. So, I had deployed overseas for a deployment and was gone for 7 months. While deployed, we stayed busy, had some incidents and operated a lot, which was good because that made time go by faster for us. We had gone to some locations that would be considered 3rd world by our standards, and part one of my culture shock was seeing small children playing with rocks, in the middle of the desert, and having fun doing so. It awakened me to how spoiled our children are back in the states. I appreciated that experience, but it was when I got back to the states where I experienced the biggest shock, and it was because of something so small and seemingly insignificant."

"Seeing kids spoiled?"

"No, it was when I attempted to use a pay phone to contact my son. I put a quarter in and tried to make the call. I was greeted with the tones that told me there was an error in making the connection."

"Disconnected line? They change the number on you?"

"No Timothy, the number was good. The problem was that I didn't have enough money in the phone to make that call... the cost was no longer a quarter, it was raised to thirty-five cents. That brought me back to reality on how life still moves forward, even if I'm not there to experience it. Does that make better sense? That was my shock."

"No, that makes perfect sense. Does everyone experience that shock that brings them back to reality?"

"Unfortunately, Tim, I don't think so. Some people never look for it or gain that wake up."

"Damn."

"Indeed. Now, go have some lunch with your mom and when you come back, we'll go over some other important information."

"Okay, Bernie, but we didn't talk about the doctors out in the civilian world."

"We will talk about that after lunch, go have some lunch with your mom and I'll still be here when you get back. Now, go eat something."

It felt as if Bernie was trying to get rid of me and if I hadn't stood up when I did, I would've missed the nurse that was standing silently in the doorway. She quickly stepped out of the room before I could see her face, since Bernie liked so much to keep his room dark all of the time, but I caught a glimpse of what seemed like fear in Bernie's eyes when I glanced at him.

"Bernie, did I interrupt something you were supposed to do?"

"What? No, I think she had the wrong room."

I tried to make a joke to lighten the mood but I could tell that Bernie was uncomfortable, so I didn't want to make it worse.

"Okay, I'm going to grab some chow. Want me to bring you back something to eat, Bernie?"

"No thanks Timothy, they have me on a strict diet in the hospital. My tray should be here shortly. I appreciate that thought, though. You go enjoy your lunch with your mom."

"Okay, Bernie. Enjoy whatever cuisine they have in store for you!"

"Ha...cuisine...I don't speak French, but unless cuisine is French for everything tastes like cardboard, it is anything but...I'm sure what you consume will be much better than what I consume."

Bernie forced a laugh as I exited the room. I could tell it was forced. This increased my concerns for Bernie.

HOW THE MEDICAL GAME IS PLAYED

CHAPTER 5

HOW THE MEDICAL GAME IS PLAYED

As I was hobbling down the hallway to find my mother, she intercepted me from out of another patient's room and greeted me with a very tired smile.

"Timmy! Everything okay?"

"Yup, lunch break. Was going to see if you wanted to sit down for some chow?"

"Sorry son, too busy, I don't take lunch at the same time every shift because of the patients. Have to feed them first and make sure all of their vitals are taken before I can sit down."

"So, no lunch with your favorite son then?"

"Don't give me a guilt trip, Timothy, I can't stop what I'm doing to have lunch with you just because you're ready to have lunch!"

"Whoa, easy mom, I was kidding with you. I know you're busy. Holy crap, you may want to eat a Snickers or something. You may be hangry!"

"Sorry Timothy, I didn't mean to snap at you."

"I get it mom, no worries. I'll go have a bite myself, so I can return for more knowledge from Bernie. He's great man...I don't understand how he doesn't ever have any visitors."

"Most of his buddies are dead and gone, son."

"I understand that but...just seems a bit sad, though."

"He has me. I'm his friend."

"You're his nurse, Ma."

"And his friend Timothy!"

"Okay, you're his nurse AND his friend. Understood. Um..."

"What is it, Timothy?"

"Can I ask you something?"

"Yes...unless it's lunch-related!"

"No, it's not lunch related...it's nurse related."

"Okay"

"Does he have another nurse on staff?"

"What do you mean?"

"Is there another nurse that takes care of Bernie...like you do?"

"You mean other than myself?"

"Exactly"

"We have 3 shifts, Tim, so he has at least 3 caretakers here"

"But how many nurses would he have on each shift? Like right now...there's you...and...?"

"And? And me...just me. Why?"

"So, each shift has one nurse assigned to each patient, right?"

"Right, but we also have aides that help out also..."

"How many for each patient, Ma?"

"Timothy, get to the point please!"

"I saw a nurse in the doorway of Bernie's room as I was getting up for lunch."

"It wasn't a nurse, I'm his only nurse on the first shift."

"That's what I thought."

"It could've been an aide, but…"

"When are aides assigned a patient?"

"Depends on the patient…but Bernie is not a patient that needs an aide. Perhaps the aide was checking in?"

"Do they do that? To patients that aren't theirs, I mean?"

"No. Not usually…unless they know them and these patients are long term, so it's not odd that some aides become friends with the patients."

"But do you know of an aide assigned to your section this morning?"

"I don't recall there being any aides on our wing…I'll check, though, but I'm pretty sure there's no aides assigned. You said it was a she?"

"It was dark in Bernie's room, too dark to really make anything out."

"But you think it was a female?"

"I'm pretty sure, unless you have a small, petite male that ties their hair up into a neat bun that is an aide at this hospital?"

"Nope, we don't have anyone that fits that description here…not that I've ever seen. I wish Bernie wouldn't keep his room so dark…I bang into things every time I walk into his room. He laughs as I curse out whatever I kicked to get to him."

"Bernie's funny."

"Yeah, he's funny and a huge pain in the butt! But he's a good man and I gladly call him a friend."

"Me too, Ma, me too. I hope he feels the same way."

"I'm sure he does, Timothy. What's not to like? You're my son, of course he's gonna like you!"

"Funny, but true…what's not to like? Anyway, I'm going to grab a bit to eat before heading back to Bernie's room for more education."

"Okay son, how's the leg feeling?"

"Hurts like it should, Mom. I'm off. Please don't forget to eat something!"

"I won't...go eat. Love you."

"Love you"

I hobbled away, pondering who it could have been in his room and wondering if they would've entered fully if I wasn't there. I don't remember what I had for lunch, only that I ate something. I only wanted to get back to Bernie's room and see if I could find out who that person was. I don't know why it bothered me so much, but it did and it obviously created concern for my mother. She tried to hide it, but just like she can spot everything I try to hide, it goes both ways and her reaction to my question about another nurse told me everything.

I made my way back up to Bernie's room and tried to act like everything was normal, or as Bernie said: my version of normal. I tried to be upbeat as I entered his room and not show any signs of being concerned about anything. He greeted me immediately upon my entrance.

"Did you have a good lunch, Timothy? How was the cuisine down in the cafeteria?"

"Bernie...I can honestly tell you that after eating hospital food, I sort of miss the chow hall on base...at least their food had taste."

"It's amazing the little things we miss and that most people will never have the experience of eating at a Marine Corps dining hall, Air Force mess hall, Army chow hall or Navy chow...it's all different in their own little way...and not all good, but special in their own ways!"

I laughed because he was correct. Each branch has their way of doing food and not every location is the same.

"That is true...I remember a lot of the instructors and company commanders would complain about the chow on base, while others bragged about the Marine Corps chow hall and their cloth table-cloths!"

"Timothy, that is a true statement. I think that was some of the best food I had eaten. The Marines truly had the best spread...but I cannot recall if I ever had chow on an Air Force base."

"I haven't, but was hoping to try each branch's mess hall at some point in time, but that's not in the cards. Oh well...can't dwell. Let's keep this going, shall we? Wait, did you eat?"

"I did eat, Timothy...I think it was some sort of sandwich...with meat? Not sure, but it was sustenance, so I can't complain...or shouldn't complain."

"I could bring you in a cheeseburger tomorrow, if you want?"

"Thank you, Timothy, but I have to stick to my diet."

"Sure, just wanted to ask."

"Thanks for asking, Timothy, I do appreciate it."

"My pleasure, Bernie...um...Bernie? Did you see that nurse or whoever she was, standing in your doorway as I was getting ready to leave for lunch? Was she a nurse or something?"

"What? I don't recall there being anyone in the doorway as you left. Did you see her while you were leaving?"

"No, as I was getting off my chair, I looked up and saw this small, petite female standing in your doorway."

"Nope, I didn't see anyone."

"I was just curious if you have more staff during the day, other than my mother."

"Not usually. That's odd, not sure who that could've been...and you say you didn't get a good look at her?"

"Correct. Just her silhouette. I was just curious if you had others that come into your room...or if I was causing a distraction and they couldn't get in to..."

"No distraction, Timothy. If there's something they need to do, they will do it...they don't care who is in the room, they'll make it happen. Most times they are not nice about it either. I don't have visitors, but some of my roommates on this wing do, and I've heard the conversations...or yelling...coming from those rooms between the patients and the staff."

"They yell?"

"Well, depending on who the staff member is, sometimes the veteran doesn't want to get jabbed again, or poked, or x-rayed or whatever and they put up a struggle."

"Staff have to get tough on you guys?"

"Sometimes...I've had to get ugly a time or two...depends on who the staff is...ANYWAY, let's talk about the medical community that you'll have to face when you have to go through the appointments for appeals or whatever the Kingdom feels you need to go through."

"Great, let me get my notebook and pen out again."

"Take your time, Timothy. I'm not going anywhere."

I sat in the recliner and placed my crutches onto the floor. Freed up, I withdrew my notebook and pen, and that was when it struck me: what if Bernie was trying to give me a hint that I needed to bring in a digital recorder and leave it in his room? I went cold for a moment, thinking that perhaps I missed the signal and Bernie's cry for help. My face got red quick from thinking that, perhaps I was overthinking things and making something out of nothing.

"Are you ready, Timothy or are you going to stare into your bag all afternoon?"

"Sorry, Bernie, no I'm ready. I was just thinking about something."

"Care to share?"

"Naw, it was silly. Anyway, we were talking about civilian medical professionals and physicals?"

"We were. Are you settled then? Ready?"

"Yes sir...sorry, yes Bernie, I am ready!"

"Good lad. Okay, so first... you should know that even the nicest of medical staff could have the worst intentions. I've had a lot of experience with varying doctors, nurses and other staff, which opened my eyes to the fact that I could only trust myself in this cold, cruel world, and a handful of teammates. That's it."

"Why did you have so many experiences? Were they all bad?"

"Timothy, I was blind and ignorant. I wanted to believe that every medical staff out there, in our world, had my best interests as a priority, and I was sadly disappointed a majority of the time. There were some great professionals, which I would recommend to my teammates who were getting out, but the majority let me down overall."

"Were they unprofessional?"

"They acted like my health was their main priority, but when I began to pay attention to things, I could see the evidence that proved otherwise. That was a slap in my face, which I needed, and woke me up to change how I viewed things at these appointments."

"Bernie, could you give me some examples so I understand the scope of what you are talking about, please?"

"Surely, but remember not every appointment ended poorly, or was conducted badly. I want to make sure you understand not all doctors were bad. In some locations it may have been a great doctor, but crappy administrators up front, or a great nursing staff, but the doctor was subpar. It varied."

"Sure, I get that..."

"Examples, right...let's see now. I'm trying to organize my thoughts in a manner that would make sense of my points, otherwise it could confuse my point."

"Whichever order you need to go in, Bernie, I'm writing it all down so I understand each point. That's the only way I'd understand the big picture, because right now I'm even more petrified to get out than ever before!"

"I'm not trying to scare you, Timothy, I'm just wanting to make sure I put it into perspective for you. Okay, so let me just rattle off an example and what was learned. That may make it clearer for you to understand. Are you good with that, Timothy?"

"That's probably best. Thanks, Bernie."

"My pleasure. Okay, the first example is expounding further on one I already mentioned when I told you about the man from the VSO that came and briefed the special warfare guys and told us that valuable information. After going through the appointment, where the doctor was trying to get me to bend over without giving me specifics, after it was over, I stopped at the front administration counter and asked for a copy of the appointment write-up. I was very distinctly told that they don't have any paperwork at that location."

"Which location was this, Bernie?"

"Oh, right, sorry. This location was an organization that contracts with the Kingdom to conduct the physical exams. The King sends veterans to these organizations because these organizations tend to write up assessments that allow the Kingdom to quickly deny claims. How do I know this?"

"I didn't ask that, Bernie. I..."

"It was rhetorical Timothy. How did I know this? I knew this because after I was told the clinic didn't have any assessments available because they send everything to the Kingdom, I then asked them how I could get a copy. Now, I've had to ask this more than once because the first woman told me she didn't know and blew me off. Luckily for me I had another appointment at the same location very soon after this one, and I was able to ask someone else who worked there. They told me that I needed to submit a Freedom of Information Act request for the specific appointment, to the Kingdom. Then, I needed to wait until it was processed before the King would send me any written assessment for that appointment. It's all a waiting game, Timothy. All a waiting game."

"Bernie, why do they take so long to process things?"

"For a couple of reasons: first, they will always outlive the veterans and second, the King knows the average veteran will forget about their claim and what happened at a specific medical appointment that occurred 18 months prior. Do you remember the specifics from an event that occurred 18 months ago? Do you remember exactly what was said, what noises were present, what you felt? Probably not, right?"

"Bernie, I barely remember what I did this morning."

"Exactly. This is also why I began bringing a digital recorder to every, single appointment. There was no way, especially after some traumatic brain injuries, that I would remember specifics at the appointments. This is also why I brought a notebook and pen to every appointment. I wanted the doctor and nurses to see that I was taking specific notes of that appointment. This helped keep the appointments on the level."

"Did any of them have a problem with you bringing out a notebook and taking notes, Bernie?"

"Not usually. A couple of doctors and nurses got a little irritated, like I had accused them of being inept or untrusting, but I just kept taking my notes. Now, back to the matter at hand: the appointment. I had a recording and notes of an appointment, so all the specifics were written down and recorded. I then submitted the FOIA request and it struck me that I might not remember sending out the FOIA request if the Kingdom were to delay a reply, so I needed to think of a way to organize everything I did. I did this mostly for my lack of memory and brain issues, but also to create a simple, big-picture tracking document for everything I was doing. Are you any good with Excel? The Microsoft Excel program, Timothy?"

"I'm pretty okay with Excel and Word. I've done some Powerpoint presentations also, but mostly worked in Word. My knowledge of Excel is minimal, but enough where I could organize the data."

"Great, Timothy, because I used Excel to organize much of my veteran stuff."

"How did you set it up?"

"I had many tabs and documents depending on the subject. I had a FOIA document on Excel that had the first column as the date sent, the second column was the date of the appointment, the third column was the topic of the document, such as TBI appointment or whatever, and the next column was for the date I had mailed the FOIA off to the Kingdom. Now, before I go on, you should know that I would send everything through the postal service and pay the extra couple dollars for signature receipt. This means someone has to sign for the envelope."

"Like certified?"

"No, Timothy, I found out the hard way that the King got wise to veterans sending their appeals and such through certified, so they changed who was eligible to sign for each envelope. This means a manager or supervisor had to sign, and there's no telling when they would get down to sign for the package. This is why signature receipt is important: anyone can sign for it. This also allows for you to track your package and screen capture the signature of who signed for it, when it was signed for and such. This becomes part of your record-keeping. This is how you do everything! Signature receipt, write that down!"

"Got it, Bernie. Signature Receipt"

"Perfect. Now that I had that document organized, I would keep a binder that had the copy of what I was sending out. I would also write the date and time it was sent, on that copy. I would staple the receipt from sending it through the mail, because on the receipt was the tracking number that you would have to enter into the USPS site, to get the screen capture of who signed for the package."

"Uh, Bernie, it seems like a lot of things to do for every appointment, but I'm assuming it was necessary?"

"Timothy dear boy? It was more than necessary. I wasn't doing this previously and was consistently appealing denied ratings for disabilities I left the Navy with. I couldn't understand it because it was in my records, it was in the hundreds of statements from my teammates, and yet the King denied, denied, denied. But back then the denials would appear up to 3 years after the appointment. Some were 2 years and other 2 1/2 years. Get my point so far?"

"That there was a delay? Sort of."

"No Timothy, not just the delay, but that I was never going to remember the appointment specifics from 3 years prior."

"Well..."

"Timothy. Okay, let me continue and it should clear things up. When I finally received the FOIA information from the appointment, I compared the appointment to my notes. For this specific appointment I only took notes, I was not recording it and that was yet another reason for me to record everything and create transcripts of the appointments. Now, reading the written assessment from this clinic and comparing what the assessment had mentioned and what I had in my notes, it was completely opposite of what actually occurred in the physical appointment."

"Did you remember any part of the appointment, Bernie?"

"Very little, but with notes, I would add small details so perhaps it would spark my memory to that specific appointment. My memory was failing miserably and I could not rely on it, which is why I did what I did. Now, I was extremely angry that the write-up did not match the appointment, but what could I do about it? Well, I used my notes to write my appeal and fought their denied rating results."

"What'd they do with the write-up you sent?"

"Timothy, I want to say they focused on screwing me better. The tactics changed and actions against me differed depending on what the appointment was and where. I also had political influence."

"What's that Bernie?"

"It's when you ask a politician for help and they contact the Kingdom on your behalf, but that's a lesson for another time."

"Got it."

"Okay, great. So, seeing the assessment opposite of what occurred in the appointment, it made me realize I needed to be even more specific in my notes, but to also ensure I submit a FOIA for every appointment, and that became my routine. I would go home, look over my notes, add additional information of how many people were inside the waiting area, how the front desk staff was, if the nurses took off running to get me to speed up my movement, etc. I would then create the FOIA request, fill out the Excel document, print two copies of the FOIA request, and put one copy in the binder and prepare the other for mailing."

"That's a lot of steps, Bernie."

"Yes, but had I known this information from the beginning, my case would not have gone a couple of decades, or three decades for some veterans. I would've had everything organized and prepared. There's no telling how many of the previous denials were because of improperly written assessments."

"Lies, Bernie, right?"

"I don't like to use the word lie, unless I could absolutely prove it. Now I skirt very closely to that term, but I don't commit to it unless the evidence is plain for all to see. It may look a certain way, but unless you can prove it is absolutely as you say it is, then your value is diminished and you don't want that to be how you're known in the veteran world. That's a quick way to never be taken seriously, and/or get a target put onto your back. Do you understand that, Timothy?"

"I do, Bernie. Sucks that you had to learn that the hard way."

"It sucks worse, Timothy, when I forgot to mail in the FOIA request or forget to log it into my tracking document immediately upon return to my home. Some of FOIA requests were never fulfilled and no reasons were ever sent."

"What? It's not an official request?"

"It is an official request, but that doesn't mean someone didn't misplace it purposely, I mean on accident, I mean on purposely-on accident."

"Oh, I get it. That sucks!"

"It does, Timothy, but it doesn't have to if you are organized and prepared from the start, which is why I am giving you this information."

"Do you have other examples? Oh, and what are the names of the organizations to look out for?"

"I do indeed, Timothy, I do indeed, but as far as the names of the organizations and clinics are concerned, I don't want to tell you because I want you to be suspicious of each one. This way, if you are sent to someone I don't mention, you are suspicious and remain on your toes, instead of thinking you're safe because it wasn't one of the ones I had mentioned. Make sense?"

"It does, sorry I asked Bernie. That's a smart way to lay it out."

"Don't be hard on yourself Timothy, I want you to ask questions, but I need you to understand the overall goal, also. Another example...Because of my hearing loss and hearing damage, which, by the way, most don't realize that is two, separate claims, but the King will not tell you that. I was told that by the contact who conducted the hearing exam. It was, again, at the clinic I was usually sent to by the King, to conduct an exam on the specific ailment. This time, however, it was conducted by someone who actually cared about veterans."

"That's good!"

"It was, but it also showed me that I still gave too much trust. Remember, this clinic was not maintaining assessments, so I would have to FOIA request the results. I did the testing and when finished, sat down with this person. I was defensive because of past experiences, and this person looked at me and verified my duty station before I was discharged. When I told this person, they brightened up and told me their spouse worked at the same Team I was assigned for those many years. Of course, I was excited, because this person continued to tell me that those who worked in the occupation I worked, had hearing damage and hearing loss for the most part. I was told that the test showed I didn't have hearing loss and I was immediately shocked."

"Bernie, I thought you had hearing loss and damage?"

"Apparently one can have hearing damage and not hearing loss. I explained to this person that I could barely hear out of one ear and mostly deaf in the other. According to the results of the hearing exam, I could still hear well enough for the results to note minimal loss and not enough to consider it as a loss. The Kingdom likes to use age as a reasoning for denying claims. I was still young at this point, and was going to appeal any denials if I had the proof in my records, which I did. I was adamantly assured by this person, of exactly what was going to be written in the assessment. I was read the report as it was being typed. I was elated that someone, finally, was going to take an interest in providing the facts in their reporting."

"But?"

"I was denied."

"After the write up that said you had hearing damage, Bernie?"

"I forgot to mail in a FOIA after that appointment because the person made me happy by telling me what was going to be written and the common link at my former Team. The denial showed up a year later, which sparked my memory about the FOIA request. One look at my logs and I realized I never submitted it. I immediately filled out the FOIA request, wrote the exact date of the examination and the location the exam took place and mailed it off. I was upset that this fell through the cracks. When I finally received the FOIA paperwork from the Kingdom and read it over, my heart sank."

"What did it say?"

"I didn't say what she had said"

"That person lied?"

"I don't actually think this person lied to me...but I do think the wording was changed by someone higher than this person. I cannot prove it, but after our discussion, I was confident that this person looked out for veterans, especially those from my community, so I don't think this person was lying to me at all, but I do believe a supervisor changed the wording to allow the King to deny me the hearing damage claim."

"Damn!"

"Damn is right, Timothy. But I'm not finished yet. What I was ignorant to, but was able to witness firsthand, was that the King will send your file to the clinic, or physician where your appointment will be. Now, I have to assume the file should contain the entire record, or at the minimum should contain every, single piece of documentation related to the specific disability being claimed. In other words, if I was going to be seen for PTSD, everything in my file which the Kingdom had in their possession, that has PTSD or trauma-based occurrences, should be in that file folder. I would also have to assume that file would be sent within enough time for the professional to take the necessary time to review the records or documents, to know exactly what the veteran was to be evaluated for."

"Let me guess...not the case, Bernie?"

"You'd be correct in that assumption. I had an appointment at the clinic where most of the appointments were, and I knew my brain wasn't working as well as it used to, and it was aggravating and embarrassing to me if I forgot someone's name or if I went blank in the midst of a conversation. To avoid feeling bad about this, I would prepare myself before each appointment, in the parking lot. I would read the date, the appointment time and the doctor's name over and over. I did this to be respectful. Apparently, it was not looked at as respectful, but instead used against me during the evaluation."

"Against you?"

"Timothy, remember there are good doctors and there are bad doctors. Sometimes the great doctors don't have a bedside manner, sometimes the doctors with great bedside manner might not be the best doctor."

"I get that."

"Good. Well, this doctor was pretty aggressive in his questioning. I answered the best I could and was spoken at with distaste the entire time. He scoffed at me, he rolled his eyes when I was trying to recollect the date and so forth. I was trying to suppress my aggravation and anger, but he was making that very hard. Upon the conclusion of the appointment, I had to sit outside in my vehicle for about 20 minutes, just to calm myself. My heart was pounding so hard, I thought I was going to pass out."

"Damn."

"I submitted the FOIA and was blown away by what that doctor had written, but remember, back then the response time of a FOIA, a rating, or a denial, was more than a year in length, so that was a lot of waiting. Anyway, I received the denial letter and the written assessment from the doctor. I was extremely aggravated again."

"The write up was bad?"

"The write up was written in such a way that lessened the actual situation. His report was filled with untruths about what I answered and even mentioned how quick I was to answer what the date was for that day. I explained to the doctor, when he immediately responded to my answer with a sudden, fake amazement on how quick I could state the date, that I spent 20 minutes in my vehicle trying to memorize the date, the time and his name. He made it sound as if I knew the answer immediately, but deliberately botched it the second time."

"So, he asked you the same question twice?"

"He did and did so with a bunch of time in between, which is why I couldn't remember the date the second time and had to use a process I created to recollect certain things. He thought I was faking everything and that made me mad."

"I'd be mad also, Bernie. That's a horrible doctor."

"Not done with him yet, Timothy. I reached out to the politician who had helped me in the past, and I informed this person of everything that happened at that appointment. Very soon I received another appointment date with the same doctor."

"Oh no, they couldn't get you to a different doctor?"

"Why would they want to, Timothy? I'm sure they choose their contractors by their willingness to write assessments that counter the veterans' disabilities and allows the Kingdom to deny the claim."

"That's evil, Bernie. How do they get away with that?"

"They are one of the largest Kingdoms in the U.S. and who's going to take the effort to hold them accountable? Lawyers? I have stories about that also, but no, this appointment was going to be different. I was prepared. I brought my notebook out and kept it in my hand the entire time. I also had my digital recorder on. I was ready for this."

"That's awesome!"

"Preparedness was awesome, but the problem is I went into that appointment looking for a fight. That's never a good idea. That's another key lesson, Timothy, don't allow emotion, as hard as it is to control, into any of your interactions or appointments. It'll only be a negative experience with negative results. It's going to suck, I ain't going to lie, but if you maintain your composure, it will help you out in the long run. Do you understand that point, Timothy? It's very important that you understand that point."

"I understand it, Bernie. I don't like it, but I understand it."

"Good. I went into the appointment and sat next to the desk. The nurse who escorted me to the room left, so I looked next to me and spotted a very large file. It had my name on the cover, so I opened it. Lying on top of the paperwork was the letter I wrote to the politician. My letter."

"But you were there for another appointment for memory or PTSD, right?"

"That is correct, Timothy. That folder should have only contained my medical documentation and any pertinent and relative information. Instead, the top letter was the letter to the politician explaining everything that was done during that appointment. That doctor read the letter where I complained about his treatment and disrespect. At this point the only thing going through my mind was that the doctor was going to be biased against me on this appointment. I closed the cover of the folder and sat in the chair, waiting for the doctor to arrive. Surprisingly, the doctor entered, sat down and proceeded to go through a shortened version of the evaluation. It was apparent he read the letter, so he calmly made sure I understood everything and he then had me listen as he recorded the assessment."

"He was recording as well? The doctor?"

"They call their service and orate the appointment specifics. Then, whoever he pays to transcribe his information into text, listens and writes it up."

"He spoke into…?"

"A phone. He called his service and basically left a message. The point is, he recorded the assessment, looked at me and asked me if what he just said was appropriate and relative. I told him yes and he pressed a button on his phone and then hung it up. He told me he would rerecord it if he said anything that was not true. I had to give him credit for actually doing his job. I never ask for anyone to lie, and I'm not willing to lie for anyone when it comes to disabilities and claims."

"You should have become a VSO or worked as one, Bernie."

"Timothy, that's yet another story down the road."

"Holy crap, Bernie. So, the doctor recorded the assessment. Did you FOIA request the write-up? Of course you did, I'm sorry, stupid question."

"Not stupid, Timothy. Not at all. The write-up matched what I heard, and I had recorded from the appointment. I left satisfied that when I did get the assessment, it would match. That was an odd feeling, and it wasn't the last time I had to go see this specific doctor."

"It got denied again?"

"They have a tendency to make things rough, especially towards a veteran who had some good firepower fighting for them. They don't like that, and you'd be forever labeled. So, the lessons there are many: try to avoid using politicians, if possible, otherwise you can be labeled and that'll make life harder on you, don't write anything that could come back to haunt you, but if you do write something, be man enough to stand behind your words. That should keep you on the straight and narrow and hopefully avoid the emotional aspect to the experiences."

"Don't get emotional. I wrote that and circled it, Bernie"

"Good, underline it also because it's an important item to memorize"

"Why did you have to see this doctor again?"

"For PTSD"

"Again? But you had to see him two other times for a PTSD rating, right?"

"Sort of, initially it was for memory loss. The Kingdom worded the appointment and the questionnaire as memory loss. I thought the second appointment was for PTSD. The memory loss and PTSD were both listed on the paperwork, but the doctor was trying to just do the memory loss side of the appointment. That's a double-edged sword, though. So, watch out about dual appointments."

"I don't know what that is?"

"This happened to me more than once...I actually had to go back years later for another PTSD exam because the King reduced my rating."

"What? Why would they do that?"

"I think it was to let me know that they had all of the power and I had little."

"That's garbage, Bernie...no, wait...that's abuse!"

"It kind of is, Timothy. But the last appointment I had for PTSD was also for memory loss and this occurred because the previous appointment was written up with a lot of lies. Now I can definitely say lies because I recorded it and typed out the transcripts, which gave me the proof I needed that these appointments were not on the level. But the recent one I was referring to; the doctor came out and told me that he couldn't do anything about the memory loss assessment because he could only do the PTSD because of his certification and experience. See, by combining the appointments and giving a doctor another tasking that they are not certified to judge, when the assessment goes to the Kingdom, it would only be for PTSD."

"What happens to the memory loss?"

"What memory loss?...Kidding, if the Kingdom doesn't get that write-up, they can deny it completely, again! So, they deliberately assigned PTSD and memory loss to this doctor, which again, was an appointment because I appealed their latest denial, knowing his certifications would only allow him to evaluate and determine a PTSD rating. Now, before I forget, I need to also add in that I did my own research on this doctor, and I did this because I needed to ensure I had some firepower behind me if I needed to appeal this decision for whatever reasons."

"Like a background check?"

"Well, I searched the name and used some locations to find out if the doctor had been arrested, and if so, what was he arrested for. I also checked to see if any malpractice suits came up."

"Did you find anything?"

"I did, but the appointment went much better than expected and he was actually a decent doctor. He admitted that he couldn't evaluate me for memory loss because that wasn't his expertise, and even mentioned many times that he was confused as to why the Kingdom would try and assign that to him. I had that recording on transcripts as well. Just trying to cover my bases."

"Why did you decide to do that, though?"

"The appointment before him was the assessment full of lies. I did a check on that doctor and found that he was sued for malpractice in the past for...wait for it...for lying!"

"Oh my, so did you say that in the appeal?"

"I didn't but it taught me to always do the research on the assigned doctor before the appointment."

"There's a lot to do Bernie, I get it. It's important, but a lot. It shouldn't be like this."

"I agree with you, Timothy. I really do and hopefully you won't need any of the information I provided to you, on what to look out for or what could possibly happen to you during an appointment, but it is always better to have the information and never need it, than it is to need it and not have it. Right?"

"Totally agree. Totally agree."

"Same thing goes with the legwork for your situation. If you put in the effort, sure you may not need it, but if you do need it, it'll save you a lot of time and aggravation in the long run. I wish I had this information before I got out, but I had to learn the hard way. You, good sir, do not. Consider this the golden ticket of sorts."

"I definitely appreciate it."

"I know you do, but I also want you to consider this information a gift...a gift to give others who may need it. There is so much information and so many pages of these sites that exist, that most get so lost that they give up. Too many veterans give up because they are tired of the beat down and don't realize that if they fought for their case a bit, they would not get screwed as much."

"Why is that?"

"What?"

"Why is it that veterans don't fight for their cases? It's their cases. I would think veterans would want to make sure everything was good for their case and not just throw in the towel."

"Timothy, it sounds easy to say but it is not an easy tasking for veterans. Imagine this: you submit a claim for nerve damage and you wait the 2 years to get denied the rating. Now, you know you have nerve damage noted in your records, but are frustrated that they denied your disability rating request. You appeal their decision. You wait the 12-36 months to get yet another denial. Are you going to re-appeal the appeal? Probably not."

"Maybe."

"Maybe, sure, but so many veterans get the beat down of denials and just throw in the towel. Some veterans fight the good fight for years, until finally they are so worn down emotionally that they settle for what their rating is. There are many veterans who get the threat of a possible reduction in rating if they appeal the findings or overall rating. Veterans have looked at that paperwork and decided it was better to keep what little they currently had, than to chance a reduction in their rating."

"They threaten veterans? That's not okay."

"It's not okay, but it's a reality. They don't come right out and say they will definitely reduce the rating, but they use proper wording so that point comes across successfully, and they were correct in that assumption. Look, Timothy, you don't have to like what I've told you, but you need to embrace the possibilities, otherwise you open yourself up for great disappointment and a lot of wasted time and emotion."

"That makes sense, Bernie, but it would be nice if there were politicians who would actually fight for the veterans and stop the abuse."

"It would be nice. Not realistic because they like to make their money each year, and gain their donations and all that nice stuff, so their level of fight is not as tough as we should be fighting."

"That really sucks! Are there no politicians in D.C. who actually has the guts enough to get laws passed and hold these people accountable?"

"Not really, Timothy. Oh, they talk a good fight when they're trying to get re-elected, but once in office, the thought of the veteran community seems to fade into oblivion...until two years later when they have to run again for office. That's when they get passionate about their fight again!"

"That sucks!"

"It does and to make it worse for you, I have more on the subject of civilian contractors and what to expect."

"Shoot."

"Another experience I had, but with a different doctor this time, was for my feet. Too many hard landings and crashes in training caused my feet to be damaged and that claim was constantly denied. This means I was constantly appealing that denial decision. I showed up and the doctor was sort of pleasant."

"Sort of? What do you mean, sort of, Bernie?"

"Well at first, she appeared friendly, but quickly showed that she was more business, which I was okay with. She went through the questions and looked at my feet. I told her my history and then I asked her if she'd had a chance to review my medical records. Her reply was that she never received them, to which my response was to ask her how she could properly evaluate the damage caused to my feet from my training injuries without seeing the copious amount of pages within my medical records. That's when I got to see her true personality."

"Did she get angry?"

"Not really angry, but aggravated that she didn't have any proof of what I told her because she never received my files from the King-dom. That's a problem because a doctor should be able to review the records to see what has been done to whichever body part, when and how often. Now, I can tell you that the Kingdom uses the fact that almost all service members avoid going to medical often because they don't want to be labeled."

"That's true."

"I know, it was true in my day as I am sure it is true today. Well, they use that fact to deny rating a disability by saying you didn't constantly go to medical for the injury or problem with that body part often, so it must be fine. Think about it, Timothy, if you were so groomed in the military to avoid going to medical unless you had to, what do you think would happen once you become a civilian?"

"Same thing, avoid going to the doctor."

"Right! So many veterans cannot dispute that because they get out without filing, then when things begin to fall apart, they decide to put in a claim. That's when they get their denial letters and their fight begins. Then, they get told they don't have enough documentation to support a rating and the veteran gets so frustrated that they just give up."

"That sucks. I seem to be saying that a lot, sorry about that Bernie."

"It's okay, Timothy. I understand and you are correct, each time, that it sucks, because it does. Anyway, back to the foot doctor, we went back and forth regarding the specifics and she agreed that the current problems with my feet are directly related to the training injuries, but when I asked her to make sure her written report to the Kingdom said the exact phrases and terms she used, she immediately stood up and nervously informed me that she couldn't and that she had to write it up as they require it. I was irritated at hearing that."

"Did it get denied again?"

"It did. So, I appealed it again. I also had to have surgery on one foot because it felt like a knife stabbing me in the foot each time I moved it. I asked the surgeon if this was also a direct result of my training injuries and he assured me that it was. I asked him if he could write up a statement so I could submit that to the Kingdom and he fearfully barked back at me that he couldn't do that. He is not allowed to write any statements for patients because he worked for the King!"

"What? So, he agreed that it was from your jump injuries while you were in special warfare, but he wouldn't write a statement that said that? That's spineless garbage, Bernie."

"Correct. So, to summarize before adding the next appointment for my feet: I was denied a rating for my feet, was sent to a doctor who didn't get my record but agreed that it was caused by my injuries, and a surgeon who also agreed that it was directly related to my injuries and they wouldn't write a statement of what their professional assessment was. My question to myself, at that point, was: if they were willing to tell me their professional opinion, why wouldn't that be the same thing they write down on paper to the King?"

"Good question. Did you get an answer?"

"Not from myself. That denial and appeal went back and forth. I was sent to another appointment and I recorded that entire appointment. I first told the doctor, who introduced herself and the resident, before starting the assessment, that I have a request and that request was that she do her job. Nothing more. Nothing less, just her job. She asked me why I would have to ask that of her and I informed her of my previous experiences with the contracted doctors used by the Kingdom. I asked her if she had received my records and if so, when?"

"Did they receive them in time?"

"They did. She told me they received them a couple weeks prior and they spent that time combing through the hundreds of pages, which tells me they received my entire record and not just the foot-related documents. Flood the doctor with everything, so they have little time to identify valuable and pertinent information related to the actual disability."

"Another tactic?"

"Yes sir. She assured me that not only will she do her job, but she will have the resident read off the report before submitting it, to ensure they covered everything and that everything that was mentioned, made it to the report. I appreciated that. The doctor then shared with me that when they stumbled across the report from the previous doctor regarding my feet, she couldn't believe what was written. She had to double check that that person was actually a physician, because it didn't make any sense to her that, on one line she agreed that the damages were this and that and caused by the impact, but the next countered that. I agreed that it was not professional, nor a good appointment overall, and she stated that in their office they use that doctor's name whenever someone goofs up on something."

"Use the doctor's name? I don't get it."

"Timothy, let's say the doctor's name was Sherry. Dr. Sherry. This doctor would call it a Dr. Sherry if someone in the office messed up, because of how messed up that doctor was in writing her report. They would call mistakes or if someone was a knucklehead, a Dr. Sherry. Get it now?"

"That's funny and sad all at the same time."

"I know. Well, I felt pretty good about that appointment when I left, however I was not going to be lulled into a false sense of achievement, because I've been down that road before and fell flat on my face. I wrote out the transcripts and saved the recording, just in case. It was good that I did because the disability was denied again. My appeal quoted the appointment and when the timestamps of when the quotes occurred. I wanted them to know I had proof of what the doctor had said, and that should have woke them up to the fact that I record everything!"

"Did it? Did it wake them up and did they do the right thing?"

"I believe they tried to combine the rating with another rating, which is another habitual action they like to do. The Kingdom will combine disabilities to overall reduce the disability rating, whether they are connected, or go together, or not. Such as if you have memory loss and sleeplessness, they may combine those with PTSD to avoid rating you for each specific disability."

"They can do that?"

"They can and they do, but most of the time they should do that, Timothy. Don't look at that as them piling on every disability into one and then giving you a lower rating for that disability. There are times when they should combine them, whether it gives you a higher percentage or not, and in some cases they do it to justify a higher rating. So, it's right most of the time, other times it's someone seeing if the veteran is paying attention."

"Holy crap, Bernie. Seriously?"

"It's ugly, right kid? It's not for the faint at heart, but if you remember what I am telling you, you can avoid becoming the victim that has to spend decades fighting for little scraps. That's why I am telling you about my experiences. It's not been a pretty ride, or easy...but I dealt with it by learning from my mistakes and ensuring I take good enough notes to not become that victim again...at least for the same tactic, anyway."

"I wasn't saying that you did anything wrong, Bernie, I..."
"I know, Timothy, I didn't think you were referring to anything I did wrong. The system is jacked, the so-called leadership in D.C. is jacked, the organizational leadership with the Kingdom is jacked, and we are mere bumps in the road for these guys. It sucks at times, but if you keep these notes and remember my experiences, the

greater your chances are to avoid the pitfalls I encountered. That's my hope, anyway."

"Mine too, Bernie. Seriously. This is a lot!"

"It is, but sadly most will forget their experience until they get their next denial letter. Then, they will spend countless hours, aggravated and frantically searching for their records and their last appeal, and so on. I want to help you avoid those situations. If you organize your paperwork, record your appointments and create the tracking document for appeals, for submissions of anything, especially your FOIA requests, then you will be leaps and bounds better than the average veteran."

"I hope so, Bernie. What else have you experienced?"

"Are you sure you want to hear more, Timothy? You haven't been discouraged enough yet?"

"Bring it on, Bernie, I await your next story with anticipation and a little ink left in the pen!"

"Okay then, I had an appointment with a doctor when I appealed their denial of breathing difficulties, and this doctor spoke very little English. Add in a language barrier and you have a recipe for greater aggravation on the veteran side of the house."

"The doctor didn't speak any English at all?"

"He spoke English, but he was Indian and his accent was so strong that I could barely make out what he was asking, AND his demeanor and mannerisms came off as disrespectful and cold. It wasn't a good appointment...actually two appointments!"

"They sent you back? What happened the first time?"

"The first time the doctor asked me what the problem was, and I explained the issues I had with breathing and fits of violent coughing. I happened to mention that it felt like I was drowning, but wasn't in water and that my throat felt as if it was closed, and I couldn't get air into my lungs."

"Sounds scary"

"Looked worse than it was, Timothy. The doctor adamantly informed me that my throat could not close. I assured him that I understood that fact, however it FELT like it was closed when I go through the violent fits. Again, he adamantly told me it was physically impossible for my throat to close, and again I reminded him that I understood that fact, but it felt like it was closed. Different than it WAS closed. He didn't get the information. He just kept repeating that it was impossible, and I was getting severely aggravated. I ended up leaving and inevitably received the denial. Again, I appealed that decision, because I was still having embarrassing coughing fits."

"Did it get denied again?"

"Another appointment had to occur first, and can you guess who that appointment was with?"

"The same doctor!"

"Yes indeed, Timothy, the same doctor. This time he apparently was sent the appeal and my statement regarding the last interaction we had, because he had a little different attitude about the issue at hand. We did, of course, go through the same conversation again regarding my throat FEELING like it closed and his insistence that my throat cannot physically close, so I figured I wasn't going to ever get that point across."

"What did he end up saying?"

"That my throat doesn't just close and he wasn't sure what it could be. That was enough for another denied rating. That was one of the longer ones I had to deal with. That hardest part, Timothy, is the waiting, but since legislation was passed, the processing time should be under 12 months for you now."

"That's good, right?"

"I thought it would be, but then I started to think about what the King was boasting about and I crunched the numbers. You see, they claimed they were processing more and more claims, a record-breaking number of claims, 1.5 million claims, but they never stated how many had appeals pending after they processed those claims. Their numbers were high AND the executives were justifying the amount of money, in the billions, they were asking for to hire more staff to process the influx of claims and appeals."

"Wait, but they admitted that they processed over 1.5 million claims and then they asked for billions more to hire more staff to process the claims? Did anyone ask them why they would need to hire more staff if their current staff correctly processed those claims?"

"My boy, you are catching on. That was my question also: how could they process, appropriately, 1.5 million claims with their current staff and then beg for billions more to be added to their annual budget to hire more staff to handle all those claims and influx of appeals. I was also wanting to know how many of those claims were returned with appeals because they didn't do their jobs correctly the first time. The executives did admit more, which was interesting because, again, the politicians didn't ask them about any of this, but they admitted that their appeal percentages went through the roof as well. So, they admit they processed 1.5 million claims in a year, then they admitted that they had to handle an increase in appeal load, more than any other year, and none of the politicians put two and two together."

"They were denying the claims in record numbers and then handling the appeals of those who still have fight inside them, right Bernie?"

"Again, on point Timothy. Yet not a single politician called them out on that situation. I was baffled and I was watching it live online. I was losing my mind. I sent the committee thousands of messages and none of them returned my messages. They ended up removing that option from the government site, apparently they didn't like to be called out on their fraud and inability to do a job they were elected to do."

"I hope I don't ever need to use a politician, Bernie. Ever!"

"I hope you don't either, Timothy. Just know that not all doctors, just like not all politicians are poor at doing their job. I have to hold onto the hope that there has to be someone in D.C. who actually cares about doing their job to help their citizens. Otherwise, without hope, we have nothing to hold onto, and that my son, is one of the reasons that some veterans end their life."

"That's sad."

"It is, but it's a reality we have to understand. Not all suicides are because of frustration, but think of it...some veterans spend decades defending their case, trying to convince one of the largest organizations in the U.S., which was created specifically to help veterans, that their disabilities are real but keep getting denied. After a while that veteran may lose hope, and at that point, when they are in constant pain, mentally or physically, it doesn't matter which, they decide that ending their pain forever seems like a better option than their current reality. That's a shame."

"They don't do anything for veterans who are on the edge?"

"They say they do. They do have doctors who could help, but how do they help when their record doesn't reflect mental health issues? There was a pilot who committed suicide on the steps of the Kingdom, while he was seated on top of his records. He left a letter explaining that he was doing this to bring focus on mental health and the many veterans who need support and acknowledgement in their cases. They stifle the actual numbers and the media pushes their agenda, which takes the focus away from the actual problem. Don't get me started! It infuriates me the level of abuse veterans are exposed to, when all they're trying to do is gain support. Now, there are some veterans that will scam the system, which screws things up for the honest veterans, but you're going to experience that type of person everywhere. That's a sad, but true, reality of life."

"That is sad, Bernie, but I know it's true. How many commits suicide each day? Do you have an estimate, Bernie?"

"Too many, Timothy. Too many. This is why I want you to help others who may need it. You'll have the information that can save others the aggravation. Pass it along. Get the information out. There's so much that can be done to help others, but no one seems to want to put in the effort. Non-profit groups can be greedy and capitalize on the ignorance and patriotism of others, yet they continue on and it helps very little, if at all. The main point is, you need to take care of your situation and be organized, then you can be the voice for others to listen to and that, in itself, could save so many lives."

"What about the hospitals and clinics?"

"What about them?"

"Are they good? Should I try and use them as my health care option?"

"Timothy, despite the hospital telling me that I couldn't be seen until the King rates my disabilities, there are some great hospitals. They are depressing, but they can be good because of the physicians and staff inside. As with anything, you can have a great hospital, but it could have crappy staff, or a crappy hospital with a top-notch staff. It happens. The main point is for you to be as prepared as possible, but to also spread what you learned to others."

"So going to the hospital is okay?"

"I go...well, I'm here, so apparently it's okay, right? I will admit, I would look around the waiting room while I was waiting for my appointment and see other veterans who were talking to each other like they meet there daily. For some veterans, that's a reality. That's their life. I used to think to myself that I could never be that veteran. It was a sad existence and made it appear as if those veterans couldn't live a normal life, or a happy life, because they relied on the Kingdom for everything. It was slightly depressing."

"Did you have to go often?"

"No, hardly ever. I avoided going to the hospital for anything. My pain was never going to go away. I was never going to heal from my injuries, and that was a fact. So, I avoided going unless I needed to do labs or I had an appointment. That way, I also didn't have to witness all of those veterans who were loudly talking about how they're in that hospital every day, or three times each week, or whatever. I avoided the sadness, I guess."

"But you helped spread what you learned, though right?"

"I tried. Some always had a one-upper to throw at me, or they were just chronically angry at their situation, so they didn't want to listen to anything anyone had to say as a solution. You can't help everyone. Help those that are willing to accept help. Some are not, some just want to be angry about their situation, and pound their hands upon the desk and scream that life is unfair. Some will listen. Pursue those people to help. You should also remember that staff at these locations could be veterans also."

"I never thought about that, I suppose there could be veterans working at veterans' hospitals and clinics."

"I watched as one guy berated a woman behind the desk. He barked at her and screamed and screamed at the top of his lungs, and kept throwing out that he was a veteran! I was putting my book away to go take this guy away from the situation, when the woman behind the counter calmly stated that she was a veteran also and she had retired after 20 years and many combat tours. It calmed the man down and he apologized. Sometimes, an angry person needs to be angry. Sometimes they also need to be managed. This is why you pay attention and take care of your business appropriately because no one knows what you are going through, only you do! Got it?"

"Yes sir...er...Bernie. I got it. That makes perfect sense."

"Good, let's call it a day and do some more tomorrow. You only have a couple days left here before you have to go back to your command, right?"

"Yeah, I only have a handful of days left on leave before I have to go back and go through the medical board. I feel more prepared now, though, thanks to you."

"It's my pleasure, Timothy. Can you do me a favor though? On your way out?"

"Sure, Bernie. What do you need?"

"Can you ask your mother to stop in here, please?"

"I will Bernie. You okay? I didn't mean to keep you talking this long, I could..."

"No, don't think it's you, son. I'm not feeling too good and need to see your mom. No big deal."

"Okay. I'll grab her on my way out."

"Thanks Timothy. I appreciate it."

"I appreciate you, Bernie. You're a great man and I am so glad I met you!"

"You're too kind, good sir, too kind. Now go enjoy your evening and I'll see you bright and early tomorrow. Okay?"

"Bright, huh? You keep it so dark in here, Bernie, I could grow mushrooms."

That was the first time I saw Bernie laugh. It looked painful for him, not because he didn't like to laugh, but because physically it looked painful for him. Him asking for my mother was a concern, so I packed up my notebook, put my backpack on my back, grabbed my crutches and hobbled out of his room. I stopped at the doorway and gave him a wave goodbye and he nodded in return. I quickly hobbled down the hallway to let my mother know Bernie's request.

At home, later that evening, I asked my mother what Bernie needed and she smiled and gently patted my shoulder and assured me that it was nothing. I didn't believe her. I know that look of separation of emotion. She uses it with patients to not get attached, and she was really skilled at it. Once that look faded, for just a moment, I could see the concern in her eyes. It wasn't for me, it was for Bernie. I felt selfish that I was taking up so much of his time, but my mother assured me that my visit, and his ability to give me information, has been the highlight of his days.

That's a sad reality, but I never took the time to think about Bernie in that fashion. He lies in bed all day, with only the television to watch, a nurse to poke at him, and the minutes ticking away on his clock. I was never told of any visitors for Bernie, and his room was barren of anything personal. It was a hospital room. Blank, dull, hospital room. The other rooms I walked by were decorated and some looked like a small apartment. It was clear that those people had others who tried to make it comfortable for them. Bernie had no one in his life. That made me quite sad.

I sat up that night and obsessed about everything he told me. I reread my notes and spoke my questions aloud, all while scribbling little notes alongside my existing ones. Any of my questions were answered on other pages, because Bernie had a way of telling me information while supporting previous points he made. I couldn't imagine Bernie didn't have anyone in his life that could come visit, especially being on that wing of the hospital. Even the other veterans had visits from veterans from other floors, so there had to be some meeting area that the veterans were able to meet up and watch movies and tell stories. I didn't see Bernie rushing to be a part of that crowd, though. Bernie seemed like a loner and I could relate to that.

I took to heart what he told me when he stressed the point of spreading the information to others. There are so many veterans who feel hopeless and don't have the inner fight to go on, and it could be solved with information if they could be reached. It's sad to think that the VSOs don't take it for action, or the nonprofit groups take money, but do not put the same attention towards the veterans they claim to help. That angered me to think that, but I know, deep down, it happens. Sickens me. I understand the frustration and aggravation part, because I want to help others, but not knowing the way how...well...that was the aggravating part. I have to think there are others out there with the same aggravation.

There has to be!

6

COLLEGE AFTER THE
MILITARY

<u>CHAPTER 6</u>

COLLEGE AFTER THE MILITARY

As I sat at the table and drank my coffee with my mother, it struck me, suddenly, that Bernie had mentioned that he had memory loss...but he was able to regurgitate all of that information to me with relative ease. I knew I had to write everything down because of the hit to the head, I forget most everything. I am actually obsessive about how many times I need to write notes on things to remember them, and place those notes in different locations. I had my phone calendar, notebook on my phone and multiple calendars in strategic locations to ensure I did not forget anything important like birthdays, medical appointments and things I needed to be reminded of. I sat there puzzled.

"Mom?"

"Yes, Tim?"

"Is Bernie okay? I noticed the bruises and heard his cough and it didn't sound very good."

"Bernie is not going to live much longer, Tim. Sorry to say. He's a good man with few friends and deserves much better treatment than he has received."

"Treatment at the hospital?"

"I don't want to discuss patient care, Tim. Let's finish our coffee and head out. I'm sure Bernie has so much more information to give you."

"He's a walking, talking advocate. He should have become a VSO!"

"He tried. Make sure you ask him about that experience, just in case you think you want to go down that road. The topic may aggravate him a little though, but not more than any other subject he knows about."

"I'll ask him, but I need to know how he's able to spit out so much information and claim he has memory loss?"

My mother laughed and sat back in her chair.

"You think he just rattled off that information from his memory?"

"Um, yeah? How else..."

"My dear, Bernie has stacks of notes and binders of information in his room. He looks over his notes almost daily to figure out what he needs to tell you the next day. He's not remembering this information, he's spitting out what he refreshed himself on before you wobble into his room!"

"I don't recall seeing him use any notes, though"

"He hides his notes underneath his blanket...plus...how often are you looking up?"

"Oh. I don't look up often. He talks so fast and with so much information, I have to scramble to write it all down."

"You should ask him about those notes, Tim. He'll get a chuckle out of that."

"I am going to ask him. I was amazed that his memory was so good! I feel like a fool!"

"Finish your coffee and let's get going."

"Mom?"

"What?"

"How much longer?"

"We leave as soon as...."

"No, how much longer for Bernie?"

My mother stood up, grabbed her mug and smiled at me before softly replying.

"Not long, now let's get going please."

As I hobbled into Bernie's room, I paid greater attention to his blanket, despite it being so dark in his room. He was keeping one of his hands underneath his top blanket. I smiled to myself and went through my usual routine of setting my stuff down, turning on the lamp and preparing my notebook.

"How are you doing this morning, Bernie?"

"On the right side of the grass, Timothy. How are you doing?"

"I'm doing okay...can I ask you a question?"

"Of course."

"How are you able to provide me with so much information when you even said so yourself that your memory was really bad?"

Bernie smiled and looked directly at me. I smiled back as I was seated back and relaxed, and I could tell he knew that I knew. He nodded and withdrew his hand, which was holding a stack of wrinkled papers with lots of writing on it. I smiled and cocked my head as he held up the papers and smiled back.

"Aha!"

"You caught me, Timothy!"

"Not really, my mom helped a little!"

"What? That woman!"

"I felt stupid that I didn't catch that from the beginning. We've spent a lot of time together, Bernie, and it never struck me...until last night! Then I felt pretty stupid!"

"You shouldn't feel stupid, Timothy, I hid the papers well and tried to keep looking at my notes while you were writing furiously! It was all about the timing, son."

"Can I ask you something, then?"

"Go ahead."

"Why tell me about your experiences and have me write them down, when you could have given me your notes and I could have copied the pages?"

"Repetition is the key, Timothy. By listening to my experiences, your retention is minimal. If I tell you my experiences, have you write the information down AND review each evening, your retention grows exponentially."

"I never thought of that...that's brilliant, Bernie. I don't want you to think that this time wasn't very valuable to me, because it was...it is, I was just curious why you wouldn't just give me the notes. It makes sense now, plus it's been my pleasure to really get to know you."

"The pleasure has been mine entirely, Timothy. I mean that!"

"I'm not impeding visitations, am I?"

"No one visits me, Timothy, except your mom and most of the time it's because she needs to jab needles into me."

"Yeah, that's my mom."

"She's a great woman, Timothy, and I hope you take good care of her. She has been my biggest supporter since I was given this room. Timothy, this is the wing of the walking dead. We are all going to die eventually, but those on this wing are going to die much sooner than later."

"I know, she told me."

"Timothy! Hey! I'm not dead yet, kid. Don't start getting all sad and stuff while I'm still sitting in front of you. I ain't ready to go yet...plus, I should have been dead many years ago, but I'm still here. Don't start piling the dirt on while I'm not even in the ground yet."

"Sorry, I was just thinking that people like you deserve to live longer and don't deserve cancer and such."

"We are all part of a bigger plan and I survived some messed up stuff, many times over, and haven't been called up to Valhalla yet!"

"Valhalla? I'm not aware of..."

"It's where warriors go when they die."

"Oh."

"Plus, I have many teammates who are awaiting my entrance. I'm sure they are all looking down wondering how it is I'm still kicking! I will see them soon enough...but for now, let's get moving on some more information. How's that sound?"

"Perfect. Sounds perfect, Bernie"

"Great, then let's get started...What's your plan when you get out?"

"I am not sure...I really wanted to be a firefighter, but I don't think that's in the cards any longer."

"Don't dwell on it, it is what it is! That means that you cannot do what you wanted to do, which you don't know that 100% yet, but let's assume that statement was true and you are not going to be able to become a firefighter in the civilian world, what else do you want to do?"

"I don't know...I really haven't thought that far ahead."

"What about college? Did you put in for the GI Bill while you were in?"

"I believe so."

"Okay, so get an education, Timothy...but really think hard about what you may want that education to be in, then start looking at possible colleges. There's a ton of online colleges now, but you want to make sure you pursue a college or university that is nationally recognized. There are many colleges that are online only and some brick-and-mortar colleges whose degrees mean nothing in this cold, cruel world."

"How do I tell the difference?"

"Are there colleges on base? At the education center?"

"I think there are a couple at least."

"Okay, walk inside and speak with them. Don't sign anything, just speak with them and then compare notes to decide which one has what you're looking for."

"Just go in and talk with them?"

"Yup, go in, sit down and just speak with them. Prior to walking in, though, take some time and check out their websites. Obviously, their sites will have the flash and beauty to gain your attention, but look at their semesters and how long they are, look at which levels of degrees they offer, because some colleges don't offer Master's level education or Doctorate level education, and you want to ensure you are going to a college that offers the subject you are interested in."

"Makes sense."

"You need to look over what they offer, then write a list of pros and cons for each topic of concern. Once you find the university that offers what you're looking for, then you can commit to them. I was going to one university that was based out of Florida. They had eight-week semesters, which I liked, had five semesters and an intercession available, which allowed me to finish my degrees in minimal time. Some colleges have semesters that are anywhere between three and six months, depending on the college, and that was too long for me."

"What did you study?"

"I liked business, Timothy, and I figured it made sense to pursue that degree program, plus the university that I went to offered all of the degree levels, to include a Doctoral program."

"Did you get your Doctorate?"

"Nope, just my Masters...I thought about getting my Doctorate, but my memory was getting so bad that it wasn't going to be worth the monetary and time investment...it's bad enough the graduate degree went to waste because of my memory."

"That sucks, sorry to hear that, Bernie."

"Me too, but again, it is what it is. Can't dwell on what you can't have or do, but embrace it and pursue what you can. You should probably remember that."

"Got it, Bernie. Tell me about your college experience."

"This has a couple parts, actually, because of the experience with the vocational option with the Kingdom, that part was the ugly part...but I began with one University, which I was happy with while I was pursuing my degree, but was curious about the field of computers. I went to another college to speak with them, since they were a technical college, but their semesters were wonky and I figured out that it was so they could capitalize on the GI Bill money from the active-duty folks, and the veteran community. I don't like greed."

"Me either."

"Good. Well, at the time, they had a degree program that was a combination of management and computers, so I collected that information and held onto it as I attending the Kingdom's vocational brief, which lasted about an hour if I recall correctly. The person doing the brief seemed okay at first, so I didn't initially get a bad vibe from him...but when we had to do the one-on-one, that was when the idiot-floodgates opened, and they opened wide!"

"Uh oh!"

"Uh-Oh is right. I made the mistake of telling him that I was in this one university, but looked into this other technical college that had a degree program where they combined management and computers, and he quickly called me a liar. He followed that up with telling me that he was not going to pay for me to go to another college. He woke the beast by calling me a liar, so I immediately went on the defensive and informed him that he could look it up himself and he, again, told me that he wasn't going to pay for me to go to another college."

"Is that normal? For them to deny a veteran from going to a college of their choosing?"

"It's not typical, no. The second time he tells me he isn't going to pay for me to go to another college, I looked at him and asked him if he, personally, was going to stroke a check that comes out of his account. He looked at me dumbfounded and I repeated the question to him. With him looking stupid, I grabbed up my bag and told him, in a not-so-nice manner, that he could go screw himself. I left the building and sat in my vehicle for about a half of an hour, fuming mad. I finally calmed down and headed home."

"So, he didn't do anything to help you with college?"

"It doesn't end there, Timothy. When I got home, I emailed the information of what I just experienced to the politician who was helping me out, and this politician rattled some cages."

"Did it work, Bernie?"

"I soon received a threatening email from that same guy that spoke about not paying for me to go to another college. Now, Timothy, I would hope you are smart enough to know that electronic anything will never, ever, die? Please tell me you know that as fact!"

"I do, for sure."

"Good, because this guy evidently missed that memo. I replied to his email, while blind copying the politician, that I would gladly meet him in the parking lot if he felt brave enough to threaten me to my face. His email was full of lies about the meeting and my version was supported by comments he made in the presentation room, in front of the other veterans, who verified exactly what I had said in my email."

"Please tell me he was fired! Please?"

"Not sure, but his boss sent me an email inviting me back to discuss college opportunities and I graciously rejected the offer. I didn't want anything to do with that program...until..."

"You went back?"

"Not to that one and it was years later, but again, another bad experience."

"What's up with these people from these programs? They're there to help veterans, right? How are they allowed to be such jackasses?"

"Good question, Timothy. The second time I spoke with the Kingdom's contact, who tried to act invested and interested in my pursuit, but turned out to just spin his wheels because he was paid good money to look the part. I had to be evaluated by an independent contact, who was certified to assess the abilities of the candidate. I showed up rolling my suitcase filled with my medical documentation, which was a waste of time."

"Why was it a waste of time?"

"He didn't even look at it. He sat there and asked questions, I answered and then he proceeded to run me through different physical challenges, which I was not told about prior. If I was told about the physical part, I could have stretched more or not worked out that day, to avoid further injuring myself or increasing my pain levels. I did what he asked, well...as much as I was capable of doing. It wasn't until I received a copy of his evaluation that I realized he was just another crappy representative hired by the Kingdom, to write what they wanted to hear."

"The assessment was wrong? Did he lie?"

"I can say, without hesitation, that it was absolutely filled with lies. This gentleman liked to misquote me and leave out vital information from the events, which would have supported my disabilities, but instead he picked and chose what I said to fill in answers, which made me look like I wasn't hurting as bad as I was. Plus, he included a statement, but didn't put it into context of our conversation, and that makes a world of difference."

"Like what? How'd he do that?"

"Well, Timothy, at one point he was asking me to carry a box with weights inside, around the room. I struggled a bit, because my back was hurting pretty bad at this point in the testing. He noted aloud that I was unable to function normally because of the weight, but then asked me if I could do the same test but with more weight. I replied that if the building was on fire, I could throw him on my back and get us both out if I needed to, because the will is stronger than the body of pain. His report stated I had volunteered that I would have no problem carrying him out of the building. No context. That made me look really bad when that report went up to the Kingdom."

"What did you do?"

"I contacted the representative at the organization used to help with resume writing and interview practice, because they were the lead for the program where I was at. I sat down with my contact and her boss and asked them to read the report. Then I explained the difference between what actually occurred and what was in the report, and asked them if this was normal for those they use to evaluate disabled veterans. They didn't have much to say to that, so I thanked them for their time and left."

"That was it?"

"That was it. It served me no good to get angrier than I already was over the same results of the program as I had previously. I just left and vowed not to go through that program ever again, because it isn't set up to actually help the veteran. Oh, they say it does, they say it is for the veteran, but it wouldn't be that hard if it truly was...or perhaps it was just that difficult for me. This is a definite possibility, Timothy, and not that far off from what the Kingdom has done to me over the years."

"So, avoid that program, got it!"

"No, I'm not telling you to avoid it, I'm telling you to learn from my experiences and show up prepared and knowing. Know that they will note everything you do, how you sit in the chair, how you answer the question, how you walk in, how you handle yourself, what you say to them, the answers you give, and so on. If you have a recording of everything, you can compare what actually happened versus what they wrote in their report. This is why I constantly push for you to get a digital recorder."

"I should have gotten one while I was here, Bernie. I could have been recording everything you say, and write it down...I didn't forget what you told me about repetition, but a recorder would have been a good idea, and I could relisten to our sessions in the evenings. Triple the chances of it absorbing into my brain."

"That would've been a good idea, but not necessary. I don't think you're going to forget too much. I trust you to take good notes."

"I try to."

"Good. Anyway, I completed the degrees I sought out to, on my own, and used my GI Bill money to do it. I chalked it up as a learning experience and kept my notes about it, and the recordings, to ensure I would never forget that this happened to me. This doesn't mean you will experience the same thing if you decide to pursue that route, but if you use that program, they could end the money you have in the GI Bill."

"What? I put money into that program!"

"I get it, but it could be possible for them to take that money from you because they would be paying for your college. You would lose all that you had in that program, but you definitely need to check on that information to see what today's requirements are. You should speak with someone who has updated information, but I'm sure the information can be found online. The Kingdom's site is endless and easy to get lost in, but keep searching and note the locations in your notes so you know which are dead-ends and which are the best locations for which data. Keep a paper copy of great link addresses, but also an Excel document so you can copy and paste the link, rather than having to type it out each time. Some addresses are really long."

"Yeah, there are some long ones. I'll keep both active so I have an alternate location in case my computer has issues."

"I would suggest that. It's to cover all bases. Remember, don't get comfortable."

"I won't, Bernie. Can you remember anything else about the program?"

"Actually, I forgot that I had mixed up something I told you...I had gone back to the program after I received the threats from that one guy."

Bernie holds up a wrinkled page with scribbling on it.

"I forgot this...I had gone back when the politician said to. This politician had spoken with a high-ranking representative at that building location and invited me back. I went back and was told that I needed to take a test."

"A test for what?"

"Ready for this, Timothy? A test to see if I knew enough about the subject I wanted to learn, before they would say it was okay to attend a college that can teach me that information."

"Wait, what? You needed to take a test...a test to see if you knew enough of the information...that you would be taught in the college you wanted to go to? I don't understand that mentality?"

"Welcome to the confused club, Timothy. That wasn't all...I not only had to take a test to see if I knew enough about the subject before I was allowed to go to the technical college to learn that information, but I also had to have a doctor sign their form recommending me to that program."

"For physical reasons? Why do you need a doctor's okay?"

"No, not that type of doctor, Timothy, a doctor of education. Like someone with a doctorate."

"What? What does that person have to do with recommending you to attend college?"

"Good, we are on the same, confused thought process. That was my thought also. I traveled an hour to take the test, which was way beyond my scope of knowledge, and then I had to debrief with one of the people there, once the test was graded. Sitting in front of this guy, he told me that the field of study I was interested in, involved a lot of moving around, carrying computers, crawling under desks, and things of that physical nature. He was trying to talk me out of that degree field. I told him I knew all of that information and what it pertained to physically, and that because of my physical disabilities, it seemed like the best alternative because it was also a management degree program."

"The same one you mentioned before, right?"

"Correct, Timothy. I held my ground. I left after he restated that I needed to get a doctor's recommendation to attend that degree program, and they gave me the name and number of the doctor they had on file. I called and made the appointment. The best part of this meeting, with this doctor, was that he was just as baffled as I was about me needing to get his permission to attend college. He even asked me why I felt I needed his recommendation to go to school. I assured him the decision to meet him was not mine, and it was a requirement bestowed upon me if I wanted to continue pursuing a degree in that field of study."

"I would have been confused as well."

"He looked at me after signing the paperwork and told me that I didn't need anyone's permission to go to school, and if I wanted to go, then it was my right to attend whichever degree program I wanted to. He scoffed at how ridiculous this process was, and was more appalled than I was, after I told him the full story. He handed me back the paperwork with his blessing and he said good luck to me as I was leaving. He was the type of person I wished was in every position within the Kingdom. But instead...we get these other clowns."

"But you didn't go to that college for computers, right? That never happened?"

"Correct, that never happened. That was the college that shortened its semesters to make more money off of active duty and veteran students, which would have reduced the overall benefits they could get at other universities. No, I stayed far away from that college...but they liked to call me daily, about 5 times a day, for months. It's easier nowadays with these smart phones. You can block a number with the smart phones, but back then you couldn't so you had to silence the ringers or just press the end button. They were relentless and it was painful at times. Eventually they stopped calling."

"That's good. Do you think it is safe for me to go to college for something, Bernie?"

"It's your choice, Timothy. If you want to learn something, you paid for it, you might as well use it. Think about what you want to do and it doesn't have to be what you previously wanted to do, that's the best part. If you want to learn coding, or anything computer-wise, the programs are out there. If you want to learn business, or become a teacher, or a biologist, or psychologist, or a doctor, the world is your oyster, kiddo."

"I'm just not sure..."

"The key thing to remember, though, is that if you don't use your GI Bill within 10 years of departing the military, you will lose it all. That being said, if you don't want to use it, don't use it. That's the best part about being an adult, Timothy, you can choose your path and you can choose whether or not to use the GI Bill. Some things are time sensitive."

"You mean with claims and such, or just education?"

"Everything Kingdom-related. Everything. Never forget that part! If you plan on appealing a denied rating, you would submit an Intent to File, which gives you a year to submit your appeal. This is for those who sit on things, or need time to collect more data in support of their claim. I always suggest you jump on the appeal immediately. Many don't have that option."

"What else is time-sensitive, Bernie?"

"Insurance."

"What? What insurance? I thought I would be covered by the Kingdom for my disabilities?"

"You should be, but I'm talking life insurance. You will be offered SGLI, which I believe stands for Service Group Life Insurance, or something like that. I can't remember. I just remember that you will want to choose the one you want forever. Until the moment of your death, and you must maintain the payments. You can have it taken out of your disability payments. Just don't let it run out! You won't be able to renew."

"I suppose that happened to you?"

"I moved and despite doing everything required to alert all of my new address, somehow they didn't get the memo and I wasn't able to get the same level of insurance as I had before. I should have asked for a verification letter that they received my new address change, but I had a lot going on at the time and it just slipped through the cracks. Just don't let that happen to you. The only other time you get to touch that life insurance option, to raise it, is if your rating is granted or raised from your appeals. Then, you will receive the paperwork again, but you won't get the same level offered to you as before. I don't know why that is."

"That sounds stupid, actually."

"I agree, but that's what it is. Same thing goes for the higher percentages of disability. If you get 90% disability, the Kingdom will send you paperwork regarding your GI Bill money. You can choose to give your GI Bill money, or whatever is left of it, to a family member or spouse, if you don't want to use it. Same thing for a 100% rating. But you cannot sit on that paperwork for long...you need to jump on that paperwork quickly and decide what you want to do, and do it!"

"I wrote that in bold, Bernie: DO NOT HESITATE!"

"Underline it, Timothy. It is vital you don't forget these things. You can lose things quickly and then you won't get another opportunity to regain control. That's a crappy feeling because you know it was your responsibility, and you had the chance to take it for action, but decided to sit on it. No one wants to be put in that situation. Many who are, live with that aggravation and regret that they dropped the ball. Avoid that by making sure you are punctual on everything that comes your way. Plus, it's good habit in work and in play, to be on time and to address possible problems immediately, rather than letting them sit and be forgotten...until it is too late."

"That makes sense, as most of this makes sense, Bernie. You know what? You should have written a book! I'm telling you, with all of these veterans out there needing help, this information would be extremely helpful to them, especially to those who just got out, or are thinking about getting out."

"I don't know anything about writing a book, Timothy, plus no one wants to hear the ramblings of a busted-up old man."

"I did and still do, Bernie. You're a wealth of knowledge of what to do and what not to do. This information is not widely known out there, but it should be."

"Well, Timothy, perhaps you should write that book then. What do you think?"

"I can't, I don't have the same experience that you've had."

"But you have my stories and you can write about your experiences to support whatever I have told you, that applies to you as well. Could be an easy way to get the word out to others."

"Maybe. I'm not much of a writer, though."

"Timothy, you will be a pro at writing if you have to start writing appeals, and statements for your friends who are leaving the military. You'll get so much practice, it has to make you better and better each time, but I hope you never need to, to be honest. I hope you don't need to write your own appeals, or your own statements, because that would mean something went terribly wrong for you."

"You're right, Bernie. I think this is really important stuff here. Not sure how welcomed the information would be out there, though. I've seen it a lot while I was in, where the information was there but no one thought they needed it, so it became something that most ignored. Then, if or when it was needed, they screamed that no one told them this or that was an option for them."

"Seems like the military today is a lot like it was back in my day. Same thing occurred, but when people start getting screwed over, for whatever reasons, others take keen notice and that starts discussions. Discussions lead to solutions and actions, to ensure no one else gets harmed because of whatever happened. That's important, otherwise you're doomed to repeat the same things over and over and over again. In the veteran world, that could cause you even more work and rework!"

"No one likes rework, Bernie. No one!"

"That's for sure. You just make sure you keep good records and jump on these matters immediately to knock them out. That way, it's one less thing to worry about doing, or forgetting to do. Again, you're lack of attention could result in losing something forever. You don't want that. I'm sure you will do your due diligence, Timothy, and tend to the things that need the attention. Double check what your actions are going to be before you pull that trigger, just like when cutting a board...measure twice, cut once!"

"Good analogy, Bernie. Perfectly put. I will not procrastinate, which is what I believe your point was in all this. It is important and I won't be that guy. Promise!"

"Again, Timothy, I hope not, for your sake."

"I promise, Bernie."

DISABILITY BIBLE

CHAPTER 7
DISABILITY BIBLE

It wasn't time for lunch yet, but my leg was hurting pretty bad, so I decided to stand and stretch. Bernie looked at me in a puzzled manner as I stood in front of him, and that look turned almost frantic.

"What's wrong, Timothy? You're not leaving yet, are you?"

"No, Bernie, I have to stretch, my leg is aching pretty bad and I've been sitting for a bit. I figured I would stretch for a second and we can keep going. Are you doing okay?"

"I'm okay, I was worried you were cutting our visit short. I have more information to give you and...well...I didn't want you to leave without me first offering it up."

"Nope, I'm not going anywhere yet, Bernie. Just going to stretch for a second. Do you need anything? I could go get you coffee or something."

"I'm okay, I have water, Timothy, but thanks for asking."

"My pleasure. Okay, that should do me for some time now. Still aching but not as bad. Let's get back at it. What's the subject you plan on discussing with me, Bernie? Bernie?"

I glance up at Bernie and see him slumped over, with his head down. For a second it scared me that Bernie just died in front of me. I used my good leg and kicked at the foot of his bed and he shook awake. I breathed a sigh of relief as he became alert.

"Wha...what's up?"

"You fell asleep, Bernie. You sure you're okay?"

"Yeah, sorry about that Timothy, I didn't get much rest last night and I guess I just nodded off. Won't happen again."

"I could come back tomorrow, Bernie. That's not a problem, I still have a day or so left here."

"No, no that's okay. I need to give you this information sooner rather than later. I'll be okay, just feel a bit silly for falling asleep while we were talking...what were we talking about?"

"I'm not sure, Bernie, I was asking you that?"

"Um...sorry...I...I think I was going to tell you about the guidelines the Kingdom uses when they assess disabilities."

"Okay...what's that called?"

"It's the 38 CFR, which stands for Code of Federal Regulations and the Kingdom uses that to assess ratings for disabilities. It changes, so you have to keep up on those changes."

I put down my pen and looked at Bernie.

"Bernie, you haven't told me why you never became a VSO...is it possible to tell me that before going into their rating manual? Please?"

"It's not an exciting story, Timothy, but probably paints a picture on possible retaliation against what they consider a trouble-maker...or should I say...a veteran who refuses to back down from their abuse."

"I would say the latter, probably."

"I was doing my research, because I didn't want to be involved with any of the VSOs any longer, especially after what I was able to figure out, and I found out that an individual could get certified through the Kingdom to be an independent agent. It involves on-going training to maintain that certification and is usually done by lawyers, since they figured out they could make a ton of cash off of veteran cases. It was a simple process of submitting an application with references."

"Sounds easy enough."

"One would think, Timothy, one would think. Well, I reached out to some buddies and asked them if they wouldn't mind being on the list of references and fielding questions from the Kingdom regarding my application, and got their approval. The instructions stated that I needed to submit their written statements along with my application. It didn't say anything about them needing to call in at any time. The instructions also stated the process from application submission to notice of approval was 12 months."

"That's not too bad, Bernie."

"It's not...if it was 12 months. What happened was: I submitted the statements and my signed application in November. By February the following year I was contacted by their lawyer and was told that whomever scanned in the applications, had done so with my signature block hanging out of the copier, because they placed the application in the incorrect way."

"How do you put a document in the incorrect way? The picture is on the copier and you'd see the paper hanging out, right?"

"Correct. You would have to have never seen or used a copier in your lifetime, to confuse how to place a document onto the screen of the scanner. Now, the only part that was hanging out was my signature, which was the most important part of that document. I was instructed to resubmit, which prompted me to quickly ensure the clock was not restarted on the 12-month turnaround. I was told that my start date was November of the previous year, when I submitted my application."

"That's good."

"Well...sort of Timothy, you see she followed up that statement by informing me that it takes more like 24 months. I quickly corrected her and told her the website said a 12-month turnaround and she scoffed at me and hung up. I received another call that my contacts never called in to provide a reference."

"What?"

"Again, I told them that the site spelled out the directions and I followed those directions to the letter. I believe they were trying to find whatever loophole they could to stop me from moving the application forward."

"Did it work?"

"Yes and no. No, I kept on watching the calendar and counting the months that went by, and by month 26 I believe, I was contacted by a Kingdom lawyer who informed me that they never received the statements for my application to be accepted. I explained, again, everything I had done and previous contact, but this time I added in that I can now successfully prove harassment and targeted abuse by the King. First, their timeline went over the 12-month period, they deliberately scanned my application incorrectly, deliberately leaving off my signature block and they refused to acknowledge the reference statements and calls from my references. I also informed them that I had copies of each person's acknowledgement, that they completed the reference interview. This is what I asked each person to do upon completion."

"Cover all bases, Bernie."

"Exactly. It wasn't long after that, that she emailed me again and stated I could schedule to take the closed book test. I responded quickly to ask why it would be a closed-book test, since the 38 CFR changes often AND is a huge document with pages of parameters? I also added the caveat that there are zero VSO's and/or agents in existence that would have the ability to memorize the entire 38 CFR, and they would have an opened 38 CFR in front of them whenever they were handling a veterans' appeal or submission. She responded that it was a close-book test and that I could either schedule it or retract my application."

"Did you retract it?"

"I had to. I had no choice, there's no chance I would have done well on that test if I couldn't thumb through the chapters. It is very expansive and very intricate with information that is specific to each area of the body. There's no way someone like me, with memory loss as bad as it is, would do well on that exam. I could've lucked out, but I would have wasted time and energy if I didn't pass the exam. There's no way I would have passed it with my memory as bad as it is."

"That sucks, Bernie. It's so unfair that they keep doing this to you."

"It is, but it taught me so much and I used that education to help others, like I'm helping you now. You can learn from my mistakes and the abuse I received by the system, to do better than me and have an easier life because of that."

"I just wish you had an easier life also, though. It sucks that you had to keep going through all of that stuff."

"It was in my cards, Timothy, don't pity me for it. Remember what I tell you and avoid the same pitfalls that I fell into. But the 38 CFR is something that you really need to become familiar with, because it does provide the criteria for each disability rating, and lets you know what to expect with what you know about your specific disabilities. Does that make sense?"

"It does. I'll have to print a copy."

"Better invest in some ink, then. It's a lot of ink! You'd be better off just saving the documents on your computer to use as a quick reference, rather than print it off and have it change in a month!"

"Got it, Bernie. That makes sense."

"Wording matters also, Timothy."

"Wording? For...?"

"Statements to support your claim, doctor's assessments, etc. Wording has to state that your disability is directly related to your service...or is more likely than not related to your service. If it doesn't say that, they will deny the claim...or they'll lowball the claim and not connect it to your service. I have disabilities that are listed throughout my records, and will received a constant denial because any written assessments do not specifically say related to service."

"Is this for real, Bernie? It has to say that?"

"It does, or it would have to fill your medical documentation within your records. Some doctors do not like to be told how to fill out their assessments, but if they can fill out a Nexus letter to support their findings, it is golden to your case."

"Bernie? I don't know what a Nexus letter is...I know a Lexus is a car, but not sure what a Nexus is?"

"A Nexus letter is a letter from the medical professional that supports your disability, and more than one Nexus letter for the same disability will only solidify your request for a rating, but is not necessary. This is only if that physician honestly assesses there is, in fact, a disability at all.

The Nexus letter requires the parameters are met. First, your disability has to have been caused by something in service, or was an issue that got worse or aggravated a condition. Next, you have to have a recent diagnosis of the same condition, and finally you have to have the medical assessment contain the first and second points."

"So, childhood allergies cannot get claimed, right?"

"Not unless something happened to you while you were in, and now that something greatly affects your quality of life, specific to that disability, but allergies would be difficult to try to claim."

"Okay."

"The medical professional would need to be specific in their wording, because there are trigger words that the Kingdom must see, to take the letter into consideration. If it says 'not likely' related to service, the Kingdom will deny you. If it says 'highly likely' or 'definitely linked' or 'directly related to'...whatever, then the Kingdom must take that into consideration and use that as their validation that the disability existed then and now."

"Okay. I think I understand."

"Timothy, remember when I spoke about that surgeon who operated on my foot and agreed that it was a direct result of my accidents in the military?"

"Yes, I do, Bernie."

"Well, if he had actually written a Nexus letter, and stated the surgery was necessary and directly related to my accidents, the Kingdom wouldn't have denied them so many times. Do you remember what he had said to me when I asked him to write a statement?"

"That he couldn't? Or was it he wouldn't?"

"Sort of both. He stated he can't write up anything because the King restricts their options as staff at the hospital. He was quick to state he couldn't, which told me someone put fear into the staff members, to deter them from actually writing supporting documents for any veteran."

"That sucks."

"It does, and I won't forget that. He could have shortened the number of years I had to suffer because of his reluctance to fill out a Nexus letter. It doesn't change the truth, because my feet will always be wrecked, and they worked just fine before the injuries."

"Do military doctors write Nexus letters?"

"I am not sure, Timothy. You may want to check on that and if they do, then you should start seeing multiple doctors and ask them each to write a Nexus letter for each of your disabilities."

"I can't just use one Nexus letter and have it contain all of the disabilities?"

"Individual Nexus letter for each disability, at least that's how it was when I got out."

"Okay, I'll stick to that, then."

"Many doctors won't want to do the extra legwork and write a Nexus for each disability you have. You may want to request that they include the terms of 'directly related' or 'more likely than not related to military service' or 'the injury you incurred was while you were on active duty'. It should be just as effective for your claims."

"Why isn't this public knowledge for every veteran out there?"

"The King doesn't want you to know. They don't want anyone to know...because that'll wreck their big bonuses at the end of the year. You wouldn't want to impede them buying a brand-new boat with their bonuses, now would you Timothy?"

"Wouldn't want to be responsible for that one, for sure."

"Good lad, their bank accounts need to grow bigger and bigger while you struggle. You can find examples of a Nexus letter online, and I suggest you find one with a template and save it. This way you can provide it to the physician when you ask them to write a Nexus letter. That would help them format it specific to what the Kingdom requires. This'll help prompt them to actually take the necessary time to fill it out correctly. I also suggest you ask for a copy of everything they write up, for your records. Then, if the Kingdom denies a rating that you know a Nexus was written up for, you can review that Nexus, make a copy and include it with the appeal. That's a decent way to counter a denial decision."

"I could burn a bunch of discs to hand out to each doctor, which could contain the format and example, and the parameters of the write-up. That should help, right?"

"That would be convenient for the physicians, sure Timothy, but that's going to cost you to buy those discs."

"Discs are cheap now, Bernie. Heck, thumb drives are super cheap now, also. I could get a bunch of thumb drives and put the formatted documents on each, and give it to the doctors. That'll make it even easier for them. I could ask them to save their Nexus onto the thumb drive and I could pick it up, or they could email me a copy of the Nexus they submit. Either way, I wouldn't be spending a bunch of money for them."

"Whichever you decide, Timothy, just double check that the documents make it successfully onto the discs or onto the drives. That way you don't waste valuable time. Log everything on your documents used to track all Kingdom correspondence and appeals. You want to track everything!"

"Got it, Bernie. Thanks!"

"Great. It's also wise to get to know the many documents that are used with the Kingdom. There are separate documents for statements, initial submission, intent to file, and so on. Each has a specific purpose."

"But if I use a VSO, why do I need to learn those forms?"

"Well, Timothy, wouldn't it be great if you could verify the correct document was used by your VSO...well before finding out that your claim is locked-in as a denial? If the VSO doesn't use the correct form, not saying they will, but if they make a mistake, that mistake could impact you for the rest of your life. It could make the difference between connecting the disability to your discharge date, or the Kingdom making the date the moment the document was submitted. That's a big deal in the grand design of things. Wouldn't you say it was important to learn those forms?"

"I stand corrected, Bernie. So does that mean VSOs make mistakes a lot?"

"We're all human, Timothy and we are all prone to mistakes. Just because a VSO makes a mistake, it doesn't mean they did it deliberately or that that person was inept at their job. It also doesn't mean that person is good at their job. I'm just saying that we all are prone to making mistakes and if you know the forms that should be used and submitted within a specified timeframe, you know what you are looking for within the VSO's actions. That's All. Not saying they are all poor at their position."

"I remember what you said before, about VSOs. Some are good, some are not. That makes perfect sense. Why did you push away from using a different VSO after you walked out on your VSO?"

"Timothy...from being a watcher for so long, I've been able to notice patterns of things, and then pick those patterns out quicker. Plus, I followed up on the promises of these VSOs, and I researched actions within the political realm. It left a bad taste in my mouth, despite being a member of one organization and its leader for a couple years. That was enough to spoil any hope I might have had regarding the VSO community."

"What were they doing? Or what did they do that you didn't like?"

"If you were to go online and search for Congressional Committee videos, you will find the hearings where these VSOs sit in front of the Committee and express their needs, which they say are the needs of the veteran community. It didn't take long for me to hear all of their statements, which are submitted prior to the hearings to ensure these politicians can remove specific speech that they may deem harmful to their paychecks…I mean their reputation!"

"They have to send in their statements before having to read them? Is the hearing live?"

"They are supposed to be, but you should know that you may get irritated and feel you wasted hours of your life when you watch these hearings. There are many people who pander to these politicians, and hearing what they consider a priority, knowing those people who are reading their statements are making a fat paycheck, and then listening to the softball questions, if any, that are tossed at these VSOs quickly aggravated me. I don't tolerate people who deliberately kiss the tails of politicians because they idolize them…or worse…they want to gain a position within one of their offices."

"Does that happen?"

"Of course it does. Some of these national officers can't shovel the money into their pockets fast enough. This is why some fight their way to the national level…they want to enjoy their retirement from the military, plus an extra six figure salary at their VSO for being an adjutant or advocate. They may only get forty thousand a year in retirement, but add in another one hundred to two hundred-and seventy-thousand-dollar salary and you are living fat, AND you don't have to break a sweat doing so."

"Quarter million-dollar annual gain? That's incredible…sounds really good to me!"

"I'm sure it does, Timothy. I would have loved to make that each year instead of clawing my way to every job opportunity, but I wouldn't sacrifice the wellbeing of my fellow veteran to do so. You can search for salaries of every VSO out there. Charity Navigator is great for non-profit groups, but you may have to just search for each VSO and their salaries to find specifics. That will show you what they make, then you can compare that number with actual efforts, other than recruiting for members, that's an effort to bring in more annual funds and doesn't benefit the veteran community necessarily."

"I heard some groups like to push recruiting all of the time."

"I'm sure they all do. Some places are smaller than others in the local posts, but as the positions climb the ladder, there's parameters they have to meet to remain off of the bad boy list in the national offices. Just because they make a fat salary, though, doesn't mean they aren't doing anything at all. What I'm saying is, look at what they are saying and then do your research to see how much of what they said, was actually done."

"I'm going to guess that I'm not going to like what I see?"

"Possibly. What I didn't like was listening to what these VSO representatives were stressing as their priorities, yet avoided real issues. That bothered me."

"What issues do you see as problems, Bernie?"

"Lots. There's veteran homelessness in almost every state, and despite that being an ongoing problem year after year, it only made media attention fairly recently. Joblessness is another. Suicide is a big one...they like to bang their fists onto the table and scream that they want to end veteran suicide, yet they don't admit why these veterans have committed suicide...or their age groups...or what they have been battling. So, each year veteran suicide gets airtime and these people know they are in front of a live camera, so they play the role of 'passionate fighter for the veteran community'! It sickens me."

"What's the suicide number now, Bernie?"

"I've seen different announcements out there, where locations are boasting that veteran suicide is down to 17 per day. They make it sound like they are successful in eradicating veteran suicide altogether, but the truth is not being published. These politicians will propose bill after bill, each year, and do nothing with them. States had to start scrambling for funding to end suicide, but they also don't admit that the Kingdom incorporates that into their annual budget, every single year!"

"If the Kingdom is getting money from the committee each year to combat veteran suicide, where's that money go?"

"Exactly, Timothy, exactly! That's a good question!"

"How do you know what bills are proposed?"

"It's public record, son. If you go online, go to Congress.gov. There you can look through past Bills, congress-specific Bills and so forth. You can also look at everything each politician has done. Everything is listed right there. I started to track these records. I tracked politicians who blatantly lied to their voters about what they were doing. It makes me mad, but I tracked them to see if I was right in what I was seeing."

"You were right? What did you see and how did you track them?"

"Again, Excel is a great program to keep organized. I tracked all of the bills on Excel."

"Do you have any printed documents, Bernie?"

"I did, somewhere, but can't say where they would be now, Timothy. Sorry about that."

"It's okay Bernie, how'd you organize these bills? Were there a lot of them?"

"Unfortunately, it took up a lot of my time. The good thing about Excel is you can tab each section however you want it organized. I tabbed one as the Congress number, so if I was tracking the 111th Congress, I would label that tab as such."

"I don't know what that means?"

"Well, I think we're in the 115th Congress now. I would tab one as 115th Congress and that page would have all of the specific information for each bill proposed in that Congress. I would have one column to write the Bills, such as HR 1. The next column had the summary of the Bill. The next column had the sponsor, such as Rep. John Smith. The next column had the party and state of the sponsor, such as R-OH for Republican from Ohio. The next column listed all of the cosponsors and the column next to that housed the date that cosponsor was added to the Bill. The next column had the date and the column next to that had the action, such as 1 Jan 1999 and the next column would have Introduced into Committee. Each column had a purpose."

"That makes sense. What about the other tabs?"

"The other tabs could be however you want to organize it. I suggest having an Excel document just for the present Congress and nothing else. You can have many tabs on one document, but you don't want to include you Kingdom-related paperwork, appeals, FOIA requests or any other information on that same document. Keep it separate from the other stuff."

"What other tab would you include for that Congress, then?"

"So...the first tab was the main tab, the next tab would be for the Senate Committee Bills. That was the tab I would update to include every veteran bill in the Senate for that timeframe. The next tab was identified as a quick cheat sheet of the Congressional Bills."

"Every Bill, Bernie?"

"Goodness no, Timothy. That'd take forever. No, I was concerned only with veteran bills, so my focus was on just those Committees."

"Why a cheat sheet tab?"

"It was an easy way to collect the data. I could filter whatever information I needed, fairly easy. The first column was the Bill number, like HR 1, the next column was the summary of the bill, which I copy and pasted from the site, the next was the sponsor's name, next was the date proposed, then I had columns that identified if it was GOP or Democrat sponsored. Then I had it broken down in totals for Democrat cosponsors and GOP cosponsors. That was followed by columns that I used to track the progress and included if it passed through the House, the Senate or made it into law. I would add notes to each block of the date it passed whichever level."

"Did you use numbers in those blocks or just write in the dates?"

"No, I used the number one. I would add the note to the block after typing the number one in the block. This way I could filter the data easier. I also included columns that would quickly identify if the Bill was vague, if it was garbage, if it was a copycat Bill and if it was not veteran related."

"Copycat bills" What does that mean?"

"By tracking these Bills, I had realized that many of these Bills had been previously proposed by others in that Congress."

"How is that allowed, Bernie?"

"Unfortunately, they change the summary title and it gets a spot on the agenda. It all depends on who the Chair person is. This is why I don't trust politicians. Not all are bad, but most of them are."

"Yeah, but why propose a copycat Bill?"

"Well, one thing, Timothy, is to fill up the Congressional agenda. This way Bills die in Committee. Another reason is to reduce the other party's chances of succeeding in their Bill, and they can reject other cosponsors to make it look like a one-party Bill. These politicians are pretty busy, so there's a good chance they would just miss those Bills being proposed."

"That doesn't make any sense...it's public record you said!"

"It is indeed Timothy, and they do it every Congress. I began tracking these Bills early on, to identify specific actions, specific advancement of certain Bills, while other Bills sat there. It was disturbing to see important Bills fall under the pile of Bills. On the Congressional site, you can easily look at every politician and their actions. It's a handy tool to use. If I want to track one specific politician, I search for that politician and it'll have all of their stats right there. You can create filters so it narrows or broadens the data."

"I guess if it's a politician who has been in office for thirty years, there would be a lot of data?"

"You are correct. The filters are easy to use. I would look at the total number of Bills they proposed and cosponsored. I would add the filter to bring up just the House and Senate Veterans Committee information and look at how many Bills they sponsored and co-sponsored. I would then look at how many of those Bills made it into law. I would look further at what those Bills were, otherwise it would just look like they got a Bill to be entered into law, but not know the substance of that Bill, or Bills."

"Why is that important?"

"Which Bill would you consider more important: a Bill that would provide advanced medical help for veterans with PTSD, or to change the name of a veteran hospital or clinic?"

"The PTSD Bill, of course!"

"Of course, but how would you know which Bill was moved through the House, the Senate and into law if you don't look deeper?"

"That's true."

"This I know, which is why I was tracking these Bills. This data was showing me which party was deliberately flooding the Committee with Bills, which party proposed the most copycat Bills, which Bills had more cosponsors, when those cosponsors jumped onto each Bill and which Bills took precedence over other Bills. I numbered every Bill, which would tell me which Bill was proposed first. This allowed me to see what the Committee was allowing to occur by pushing some Bills, while ignoring others."

"I have to admit, Bernie, I don't know how that all works with the Bills and the Senate thing."

"Most don't and I had to look into it to understand it myself. Here's how I understand it: A Bill can be proposed in the House or the Senate. If a Bill is proposed in the House, it has to be voted on to make it to the Senate, which is where it has to go, to be voted on also. It has to pass the House before it ever makes it to the Senate. The same thing goes for Bills proposed in the Senate: It has to pass the Senate before it can go to the House. When a Bill passes one of these levels, if there are any changes to that Bill, it has to go back to where it originated and it must start over."

"What?"

"If Representative John Smith proposes HR 1, which is to re-name a clinic to the John Smith Clinic, and it passes the House votes, then it moves onto the Senate and is placed on the agenda. When the Senate Veteran Committee brings it up, say Senator Joanne Jones wants to change a part of the Bill, let's say she wants the Bill to change the name of the clinic to the John and Jane Smith Clinic, then that Bill must go back to the House and begin again. The House has to vote all over again, to pass that Bill through the House and onto the Senate."

"Holy crap, Bernie! So...any politician could hijack any Bill they wanted to? That's dark, Bernie, real dark! What would stop a politician from changing a Bill, then proposing the same Bill and only let their party in as cosponsors? Is that what the copycat Bills are about? Please tell me that's not a thing?"

"I wish I could tell you that, Timothy, but the fact is, that does happen. It is a common thing. If a politician wanted to make a change to a Bill, to add something, they could easily propose an amendment for that Bill, to make it a separate vote. That way it wouldn't hijack the entire Bill if the Amendment wasn't passed or if it was flawed."

"How come no one calls out these politicians?"

"Who do you think is paying attention, Timothy? You didn't know any of this was a real scenario until I just told you, right? Think about the average person in our world today. How many people want to spend their time and their effort to track these Bills?"

"I suppose not many, huh?"

"I could safely say that the number of people who track these things are very, very small. I understand why, though, I mean...I started doing it because it didn't make sense to me that all of these politicians were screaming about how proactive they were about veteran issues, how much they supported the veteran community and the massive number of Bills they were involved with, and yet veterans were still suffering as nothing happened for them."

"That's incentive enough to start tracking these politicians, Bernie!"

"Which is why I started. Let me give you an example, would that help explain things? I have the paperwork here for a politician I was tracking in the 114th Congress, so only a couple years ago."

"That'll help a lot, Bernie. Just so I can grasp the overall picture of things."

"I get it, I've been feeding you a lot of information the past couple of days. It can all begin to run together and make a mess of the picture you have created in your head. Okay, I was tracking the 114th Congress, which had 442 Veteran Bills. The dates for this Congress were 2015 to December 2016. Out of those 442 Bills, only 65 passed one chamber. Only 24 passed both chambers and made it into law. That would be a 5% pass rate. At this point, who's going to think about the other Bills that went away?"

"No one would!"

"Right! Out of the 24 veteran Bills that made it into law, how many were just to rename a hospital or clinic, do you think?"

"One?"

"You are not correct, good sir. 7 Bills were to rename clinics or hospitals. That brings the number of possible Bills that might help veterans down to 17 Bills."

"Crap!"

"Wait, there's more. Out of those 17 Bills, 1 Bill was specifically for Cost-of-Living Allowance, which is proposed every year to adjust for inflation, so now the number is 16 Bills. I'm not done, though, of those 16 Bills, 1 was for construction, 1 was for land, 1 was to extend a deadline of a report, 1 was an LA leasing act, and 1 was to acquire a memorial in France. That brings the number of veteran-helping Bills down to...?"

"Eleven."

"Eleven! More than half of the Bills that made it into law, were not helping veterans or their families. That brings the percentage to 2.5. Don't you think there were other Bills during this Congress that might be more important than renaming a clinic or hospital? I can guarantee there were...a lot more, actually."

"Which politician were you tracking?"

"I was tracking many, actually, but one in particular was a female Democrat from Florida, but I won't mention her name. She had been in office since 1993 and her numbers were bad. Overall, not just veteran Bills, she sponsored only 24 Bills total during her time in office. She was collecting a six-figure salary each year, not including any donations or gifts she accepted during her tenure, and she only sponsored twenty-four Bills overall. That didn't sit right with me. She cosponsored 587 Bills overall. Now, let's see what her veteran numbers were with those Bills. She proposed 2 Bills in 2015, neither made it into law and only passed the House, and one of those Bills was for budget reform.”

“What does that have to do with veterans, Bernie?”

“Nothing. That’s the point. She cosponsored 4 veteran-related Bills in the 114th Congress, and 2 of those Bills made it into law: an ID card Bill and the deadline extension of a final report. 24 years in office and those were her stats. Not impressive at all!”

“How did she get elected over and over again?”

“No one checks on these politicians and their false promises. She might have used statistics overall to boost her numbers. She might have said she cosponsored over 83 veteran Bills, but left out the tidbit that the 83 Bills were since 1993, and not her recent tour.”

“That’s deceitful, Bernie! Do they all do that? Do they all make false statements to get re-elected?”

“I wish I could tell you no. I won’t lie to you, Timothy, but you should yearn for the truth and the truth is that everyone lies. Politicians lie with style to get re-elected. If a politician were to tell the crowds that they proposed 2 veteran Bills in the House and both Bills died in Committee, but then they tell the crowd they cosponsored 30 other veterans Bills, which only had 1 Bill make it into law, then their stats would paint an ugly picture of failure.”

“But it’s the truth, Bernie. They should all tell the truth...but I’d be an idiot to think that any of these people would tell the truth...just sucks that they are so dishonest. What do they do all year, then, because it’s obvious they don’t fight too hard for the Bills they propose?”

“They propose other Bills. They are supposed to be there, in D.C., to fight for their constituents in their district and state. This means many other Bills that are not veteran-related get proposed by these people also. They barter with each other for votes on Bills they want to move ahead, be it for monetary gain from businesses or big organizations, or because they know their district constituents won't vote them back in if they don't show progress on their behalf. It's ugly all the way around."

"I don't think I could do that...or would do that, actually."

"Good. It's also why I never chose to run for office. I won't sell anyone out for monetary gain, which means I wouldn't make a decent politician, but don't completely push that option out of your future, Timothy. Who knows? You might be the perfect politician who makes changes that actually do good things for your fellow veterans. Just saying!"

"Maybe...I just don't see it now."

"That's okay. You're still young, Timothy and you have your whole life ahead of you. Now, before I forget, I wanted to remember to tell you about what you're looking at when you thumb through the 38 CFR."

"That's right, sorry if I got us off track, Bernie!"

"I think I derailed us, but this is something I wanted to make sure I mentioned in regards to reading the 38 CFR. There are many sites where you can read through the varying information about each system. There are categories that break down the ratings in more detail. This is important because it shows you where your disability should fall on the scale."

"For example?"

"For example, 4.71a Rating schedule is for the musculoskeletal system and the 5000 code is for Osteomyelitis, acute, subacute or chronic. Then, below that lies the different levels for the ratings. I have a printed screenshot of this category and it shows if you had frequent episodes, with constitutional symptoms, that could receive a rating of 60 percent. If you scroll down the page, 5002 is for multi-joint arthritis, except post-traumatic and gout, in 2 or more joints, as an active process. Then, if you have one or two exacerbations a year in a well-established diagnosis it could receive a rating of 20 percent. With constitutional manifestations associated with active joint involvement, totally incapacitating, it would receive 100 percent."

"How would I know where I fall within those levels?"

"Well, as with my case, you may have a difficult time guessing what the Kingdom is going to rate your disabilities. You may feel the disability deserves more, and you may be right, but the King may think otherwise...and usually does. You need to understand how to calculate your percentages, also. Most veterans get frustrated at the King's math, because it is not one plus one equals two, math. It's more like ten plus ten, plus ten, plus ten, plus ten equals twenty."

"What?"

"Yes, Timothy, it's not going to make sense, but I'll map it out for you to make it easier to understand."

"Great, thank you Bernie."

"My pleasure. Anyway, search online for code of federal regulations and there you can look up title 38. You want to begin on chapter 1, part 4, which is the schedule for rating disabilities, if you want to see the varying levels of each disability and each section they rate for. You'll see it listed out as: The Respiratory System, The Digestive System, The Skin, Mental Disorders, and so on. It even includes Dental Conditions."

"I'll look at that site tonight."

"Perfect. So let me go over how to read the percentage chart for overall percentage. First off, you need to organize your disability ratings when you get them, from lowest to highest. The combined rating chart can be found online also, but it doesn't explain how to use it and many veterans have gotten very upset in trying to figure it out. Once you learn, Timothy, it'll be a snap to figure out your disability ratings."

"Should I print it out and have it in my binder, Bernie?"

"You could. I just made a tab on my Excel document and made my own ratings table. It's the same table, just made it much easier to calculate my own ratings without having to rely on the internet."

"Got it."

"Okay, let's give you some example disabilities with example ratings. How many disabilities do you want to use?"

"I don't know...whatever you think, I guess."

"Okay, let's use 6 disabilities. Let's use: Left Foot with a 10% rating, Right hand with a 10% rating. Ringing in the ears, also called Tinnitus and that is rated at 10%, which is the maximum rating for that. Painful scar rated at 10%. Grinding of the teeth rated at 10%."

"Grinding of the teeth? Is that such a thing?"

"It is, Timothy, it's called TMJ. Let's see, one more...let's use tendonitis in the left knee at 20%. That should be six disabilities and the ratings, from lowest to highest would be: 10, 10, 10, 10, 10 and 20 percent, right? Now, what does that add up to, using the math you know, Timothy?"

"That would equal 70, right? Yeah, 70 percent, Bernie."

"Wrong. Using the King's math, it would be 50 percent overall."

"Okay, you're going to have to teach me how to read it then. That math is wonky!"

"Oh, it's wonky all right. When you look at the combined ratings chart, you will see columns and rows. The first column, on the left, begins with the number 19. That is the number you get when you add a 10% rating and a 10% rating together. Now, you need to re-member that your calculations must begin with this highest rating. That's two ratings so far, we have 3 more tens to incorporate and a twenty. Since the first two tens brings us to 19, we go to the 19 in the left column and then we use the 3rd ten percent rating. The top row goes by tens and begins with ten, then twenty, then thirty, etc. until you reach one hundred on the far-right column."

"So, the left column goes by tens also?"

"No, sorry Timothy, it goes by ones."

"Why does it go by ones if the top row goes by tens? That doesn't make any sense."

"I'm sure it doesn't make much sense to most. The maximum percentage a veteran can achieve is one hundred percent, but this system rounds down and rounds up. As the overall rating gets closer to that one hundred, each ten percent rating will only move that number up by one. I know it sounds confusing, but hang in there. Let's go over the percentages again, really quick: there are five-ten percent ratings and one-twenty percent rating for six disabilities, right?"

"Right. That much I get."

"Great. We know that we start with the highest percentages first, so the twenty percent rating starts it off, then the five-ten percent ratings are used. Right?"

"Got that. So far so good."

"I want to make sure I didn't confuse you by first stating the 19 on the chart. You must begin with your highest rating and then work down from there."

"But the 19...?"

"Timothy, the 19 is used if you only have ten percent ratings. Taking two of the ten percent ratings, which you would think equals twenty percent, it actually equals the 19 on the chart. So only if all of our ratings were ten percent ratings..."

"Then the first two-ten percent ratings will equal a nineteen, which is the first number we would start with."

"Does that mean a veteran has to have at least two-ten percent ratings to get a disability check?"

"No sir, a veteran gets a disability check if they have a ten percent rating. It's not much, but it's still something. Also, before I forget...and you are not going to like this..."

"What is it? Are there symbols I need to know and maybe some interpretive dance I need to do before I can calculate my percentages?"

"That's funny. It's not that severe...but you need to know that zero is also a rating percentage."

"What? A zero? How does that factor into the overall rating percentages of the disabilities a veteran has?"

"It doesn't factor into the rating numbers, however, a zero is still a rating. That means the Kingdom will take care of that disability, despite it being minor as far as ratings are concern, but a veteran could ask for an increase in rating percentage for that rating later on down the road."

"Why bother?"

"Well...it could become a big deal down the road and if you didn't claim in on your initial claim, you will pay hell in trying to gain a rating at all for that disability. This is why I stress to you to make sure you claim all of your disabilities, even if some of those problems are not causing you a lot of grief at this very moment."

"Okay. I guess that makes sense."

"It makes a lot of sense, believe you-me, and you don't want to fight that fight later on down the road. The VSO guy I mentioned previously had said to us that we are not thinking about this moment right here, right now. We are thinking about ourselves twenty years from now. That is who we were trying to help."

"Okay, it's important. I highlighted it and underlined it. Does it really matter which order we calculate things, Bernie? I mean really?"

"Timothy, it does. Let's calculate it both ways and look at the results, okay?

"Sure."

"Okay, so...using our 10,10,10,10,10 & 20 percentages. If we went from low to high, it would be as follows: two tens equaled 19 on the chart. We know a ten percent rating is next, which is our third ten, so we look at the 10 on the chart, next to the 19. That'll give us 27. Once we have that 27, we can now move down the far-left column until we reach the number 27. Our number four rating was another ten percent, so we look at the 10 percent column next to the 27 and it's a 34. Now we move down the left column to the 34 and use our fifth ten percent by looking at the number next to the 34, in the 10 percent column, which is a 41. We have a 20 percent rating left over. Now, we move down to the 41 and go down the 20 percent column to get the number 53. If we only had six-ten percent ratings, the number would be a 47. A 45 and above goes up to the next highest number. This means either way, it would be a 50 percent rating, overall."

"Wait, it would be 50 percent overall, if I had a 10 percent rating or a 20 percent rating?"

"Correct. The 20 percent rating, when using 41 as the sum of our fifth-ten percent rating, would be 53 and you would round down to 50 because 4 and below rounds down to the lowest ten."

"And that's if we start with the lowest percentage first and then worked up. But you said that's not how the calculations are equated, correct Bernie?"

"That's correct, Timothy. Always begin with the highest percentage and work down. Let's see what comes our of our calculations when we begin with the twenty percent rating first. So our numbers are: 20,10,10,10,10 & 10 percent, right?"

"Got it."

"We go directly to the 20 on the chart...."

"Wait, I thought the 19 was the first number, Bernie?"

"It is."

"I'm lost...I don't..."

"Timothy...if you only have ten percent ratings, that when you'd begin with the 19 on the chart and only if you had at least two ten percent ratings. Since we are starting with twenty percent rating, which is the highest in this example, we can go directly to the 20 on the chart. Does that clear things up a little for you?"

"That makes better sense now, sorry about that Bernie."

"It's confusing, Timothy, it is and I suppose that's what creates a lot of aggravation for many veterans. Let's get through this example, okay?"

"Okay"

"Great. So we begin by going to the 20 on the chart. Now we have five-ten percent ratings to deal with. Start at the 20 on the chart and look down the 10 column to get the number 28 on the chart. That takes care of the first ten percent. We then move to the 28 on the first column and look down the 10 column to use our second ten percent rating, which gives us 35. Move to the number 35 on the left column and down the 10 column again to use our third ten percent rating and we get the number 42. To recap thus far, we used our twenty percent and three-ten percent ratings. We have two more ten percent ratings to calculate. Now we move to the number 42 on the chart and look down the 10 column to get the number 48. Move down the left column to the number 48 and look down the 10 column to use up our last percentage of 53. We then round down to 50 percent."

"That's the same number as before, Bernie!"

"In that example, yes. But that is not how it always works out, which is why we always begin with the highest rating."

"Wow…I don't know what to say, Bernie, other than…what the hell, man?"

"I know, Timothy, I know, but once you do it a couple of times, you'll get the hang of it. Do you want to try another example? To try and lock in how the ratings will flow again?"

"Actually, I would like to go through it a couple more times, is that okay? I just want to make sure I understand it fully."

"No problem, Timothy, but I am going to make it easier for you and instead of using body parts, we can just use letters. Okay?"

"Okay."

"It'll still work the same way. Let's use this for an example: A is 20, B is 30, C is 10, D is 10 and E is 20. What do we do first, Timothy?"

"Organize them by the highest first, down to the lowest, right?"

"Correct. So, what does that look like?"

"That would be B 30, E 20, A 20, D 10 and C 10, right? It would be reverse that if we went from lowest to highest!"

"Good. That's right. Let's do this both ways also, just to prove how different it could be? Start with lowest to highest so we begin with the wrong number as a sum."

"Okay, so the wrong way is C 10, D 10, A 20, E 20 and B 30. Then the two-tens would equal 19 and 19 is the first block on the chart here."

"Correct. So those two tens are done for, right?"

"Right. That leaves two-twenties and a thirty."

"Good. What's next?"

"Okay, we're at 19 and I need to apply the first 20 percent. I move across the row from the 19 and look at the number under the 20 percent column."

"And what is that number?"

"It's a 35."

"Keep going."

"Okay, once I have the 35...I forgot what I do with that 35."

"That 35 is your sum of the two-tens and the twenty. Find that sum on the far-left column."

"That's right, we take the sum and find it on the left column, and that's why the left column goes by ones instead of tens. Okay, so I find the 35 and apply the final 20 percent. I look across the row, into the 20 percent column and the number is 48. I have a 30 percent left. I find the 48 on the far, left column and move across to find the number in the final 30 percent column, since that's my last rating percentage."

"Good and what is that number?"

"It's a 64...which means it's a 60% rating overall. Did I do that right?"

"Okay, now remember that and let's do it the correct way!"

"Correct way, highest to lowest. So... B 30, E 20, A 20, D 10 and C 10. Start with the 30 on the left, then look down the 20 column to get 44. Move down to the 44 on the left column and look down the 20 column again to get 55. Move down to the 55 on the left column and look down the 10 column to get 60. Move down to the 60 and look down the 10 column again to get 64. This means an overall 60 percent rating again."

"Sounds right to me, Timothy. Good job! So even though the math would come out to a sum of 90, using regular math, the Kingdom's math shows that it becomes a 60% rating overall."

"I think I get it. It sort of makes sense...well...it doesn't make sense, but I understand it better, is a better way to put it."

"Excellent. Do you want to try one more? Maybe a harder one with a lot of ratings?"

"Okay...it will help to see if I can do it without any help. Okay, let's do it, Bernie. I'm ready!"

"Well, you don't need to do it the wrong way this time. Let's do this the right way, highest to lowest. Ready? All right, here we go: A 30, B 30, C 10, D 10, E 10, F 20, G 10, H 0, I 0, J 10, K 10 and L is 0. Go!"

"Okay, organize first: 30, 30, 20, 10, 10, 10, 10, 10 and 10 percent. Normal math would equal 140, but this isn't normal math...so begin with 30 on the left. Look down the 30 column and get 51. Move to 51 on the left column and look down the 20 column to get 61 and move to 61 on the left. Look down the 10 column to get 65, move to 65 and look down the 10 column to get 69. Move down to the 69 column and look down the 10 column to get 72. Move to 72 and look down the 10 column to get 75. Move to 75 and look down the 10 column to get 78. Move to 78 and look down the 10 column for the final time to get 80 percent."

"Good. That was good. I like that you talked yourself through the process. It'll make it easier and easier to go through it when you are trying to gain the overall percentage information."

"I cannot believe that it only equals 80% when it's 140% overall using real math."

"I know, Timothy, I know. I've heard that same reply over and over again from others. It is what it is and that's what the Kingdom uses to get the overall rating percentages for each veteran."

"I feel a bit more comfortable with it now, Bernie. Thanks for running through it a couple times with me. I appreciate it."

"My pleasure, Timothy."

"I'm going to make my own table using Excel and am going to have a printed chart also. Anything I can do to get the information quicker and easier would be helpful, from everything we've talked about."

"Easier is better, for sure. I think, though, we should call it a day, young sir. I'm feeling a bit tired and should probably get some rest."

"Sure thing, Bernie. Are you doing okay, though? I'm not making you feel bad, am I?"

"No, son. I'm just tired. You have a good night and I'll see you tomorrow, okay?"

"Yes sir, Bernie. I'll see you first thing in the morning, if that's still okay with you?"

"It's fine, Timothy. You have a good night. Pamper your mom a bit, she deserves it after having to deal with me for so long."

"I will, Bernie. Thanks again. Get some sleep."

I packed up my notebook quickly and shut off the light. I could tell Bernie wasn't feeling his best, but it seems he quickly got worse. It concerned me, but I didn't want to pressure him or question him to find out what was wrong. I did what a respectful person would do, I left for the day so he could get some sleep.

ORGANIZATION

CHAPTER 8

ORGANIZATION

The next morning, as I was trying to organize my notes over coffee, I recognized that my mother was unusually quiet. I made an attempt at small talk, but she would just nod and smile. I wasn't going to get my mother to talk about anything she didn't want to talk about, but I made an attempt to use Bernie as the subject, to get her to talk a little.

"Mom? How much time does Bernie have? Is he sick-sick?"

"Um, what Tim? Is Bernie what?"

"How sick is Bernie?"

"He's dying, Tim...so he's pretty sick. Why?"

"How much time does he have?"

"It's never enough, Tim. No one knows for sure."

"The doctor doesn't know?"

"The doctors can only give estimations from the data provided. Bernie's paperwork only gives a snapshot to the doctor, and the doctors can only go by what their books and their past experiences tell them. It could be a day, or it could be a month...or even another year."

"A year? Really? Do you think...?"

"I don't, Timmy, sweetheart, I don't."

"Can I ask you another question about Bernie?"

"If you must, son."

"How is he getting those bruises?"

"He tells me that he knocked into things...in his room."

"Do you believe him?"

"I think Bernie would lie to me if he felt it would save me from certain information."

"If he's not doing it to himself, by bumping into things, how else could he be getting the bruises...and the bloody nose?"

I stared at my mother, waiting for an answer and she just sat there and sipped her coffee with glassy-eyes. We stared at each other for a couple more moments before I turned my attention back to my notes. I attempted to lighten the moment by complaining about how messy my handwriting was and how I must have inherited that from her, but she wasn't really paying attention as she sipped at her coffee. She was concerned about something, specifically about Bernie and his situation, and was thinking really hard about something.

"Mom?"

"Yes dear, what is it?"

"When are we going to head to the hospital? I know you don't work today, so you can just drop me off. I can call you when I'm ready to come back. I have to get my clothes together anyways tonight, so everything is ready for me to head back to base."

"Time goes by so fast, Timmy."

"I know, Mom, I know."

"We can leave whenever you're ready."

"Okay, I'm gonna get my stuff together and I'll be out in a couple minutes then. Thanks for everything mom, for introducing me to Bernie, for letting me stay here with you. For everything!"

"You're welcome, son. Go get your stuff ready."

I stood up from the table, pulled my papers together quickly and stepped away while my mom remained, still sipping at her coffee. I wasn't going to get any more information from her and probably not going to get information from Bernie either. At this point I felt stuck, unable to help either of them. It was aggravating at best, but it seemed to make Bernie happy to talk about everything he learned throughout the years. My goal this visit was going to be making him understand exactly how much I appreciated everything.

I had little time to think about things as I entered Bernie's room. I smiled at the consistency Bernie held to keeping his room so cave-like. I greeted Bernie as I had all week, but only received a slightly raised hand from his bed. I ignored the odd greeting and a went through what became my routine each day I visited here: Set my bag down, turn on the lone light next to the recliner, sit down and retrieve my notebook and pen.

"Okay, Bernie, what golden information do you have for me today?"

I could tell Bernie wasn't feeling good because up until this point he hardly moved, and I happened to notice his shade was completely drawn, rather than partially. It was obvious he wanted to remain in the dark.

"Are you feeling okay this morning, Bernie?"

"I'm good, Timothy. How are you doing? How's your leg feeling today?"

"It aches, but I suppose I should get used to that."

Next to Bernie's bed was a stack of three boxes, each stacked atop the other. Bernie's name was on the side and the boxes looked old.

"What's in the boxes there, Bernie?"

"A little gift for you, Timothy. Nothing special."

"Gift for me, Bernie, you didn't need to give me anything. This information is more than enough! I still cannot thank you enough!"

"These boxes contain the past three decades of my battle against the Kingdom. I think there may be some items inside that could help you out."

"But Bernie...that's your medical information. Don't you want to keep those?"

"What good will they do me now, Timothy? Come on, son, I know I don't have much time left on this planet, I think you perhaps you should embrace that also."

"I'd rather not, thank you."

"Timothy, it's how life works. If anything in these boxes help you, in any manner, then the work it took to compile this information was worth it. Otherwise, it's all for not and I'd rather leave this Earth knowing it helped someone out."

"Bernie, the information you gave me is going to help a lot of people. You should embrace that fact, if I have to embrace the other thing."

"Fair enough."

"What do you want to talk about today?"

"Timothy, I want to touch on the organization of your paperwork again. I think it's important enough to go over it again, to make sure you understand how important it is to stay organized and not let your guard down. Is that okay with you?"

"Yes sir, uh...yes Bernie."

"Good. These boxes have everything I organized myself. As time progresses, though, I'm sure you may need to add sections or information, to apply specifically to your case and whatever changes that are made with the laws and the way the Kingdom does business.

Basically, I'm telling you to stay on top of this information, but don't let it run or rule your entire life. Organization means you can have a life, and enjoy it, without having to spend more time than necessary on your veteran-related case."

"It's important, I do get that Bernie. Let's talk about it. Where do we start?"

"Binders, let's begin with binders. I began collecting my information into one binder and used separators to divide each category, but that filled a binder quickly and caused me to invest in another binder, and so on. It made it harder to keep things truly organized in a manner that would have given me the best chance at locating the information I needed, in a timely manner."

"How many binders do you think I should get?"

"None."

"But you just said…"

"Timothy, I'm giving you these boxes to take with you. I've asked one of the maintenance fellows to bring them down to your mom's car when you leave. There are binders inside these boxes that you can use as you see fit in your organization-related pursuits. You can shred my documents or toss them into a fire. It doesn't matter to me what you do with my medical paperwork."

"Okay, thanks…Bernie."

"Those big binders get pricy, so it saves you a little."

"Okay, how to organize them, then."

"With the past decades used as a template of what sort of information is needed and collected, I would use one binder just for my medical documentation from the military. I would use post-it notes to tab each page for what the disability was or injury location. I did this with mine, but back then there were very few color choices for those post-it notes. I had the big ones and cut them into smaller ones and tabbed the pages with those. For example, I would label my foot injuries as 'Feet' and every page in my records that related to my feet issues, had a tab located at the top of the page and in the same spot."

"The same spot?"

"Yes, so I could look at the tab and find 'feet', then I could look at every other tab that was located in line with the 'feet' tab. That way I could arrange my post-it notes according to the location of the injury or the body part and quickly identify the next one that contained information for my feet."

"Okay, that makes more sense."

"One binder for my medical records and have each page tabbed. It will take you some time, but it's worth it in the end. The best part is you won't have to waste time to thumb though all of the pages to find the pertinent information. That takes care of one binder and you could look at your medical records to determine the smallest binder to use."

"You have different sized binders in there?"

"I do. Use the smallest one you can for your military medical documentation because that file won't grow once you get out."

"That's true. Okay, small one for my military medical documentation."

"I'd suggest having a copy of your service record and your dental records also. I don't remember if I told you that before, but it's a good idea to have those and make copies also."

"I have that note in here somewhere, but will write it down again, Bernie. It's important."

"So many veterans get aggravated because they left the military thinking the Kingdom would be this magical organization that helps veterans navigate life once they become a civilian, then realize the whole idea of that was so foolish, and they never thought to make copies of their service and dental records. Some never thought to make copies of their medical records. That's silly to me, but then again, I've been battling this war for decades, so most things may seem silly to me."

"I will definitely make copies of my service and dental records. I don't know where or how I would get a copy when I left the military anyways. Better to have it and not need it, right?"

"You listened well, Timothy! Good job, son. The next binder I would use for the Kingdom-related disability claims and appeals. If you use a big one, you could add in a section for FOIA requests also, but with the quantity of requests I submitted, it's probably wiser to have a binder just for FOIA requests and a tracking document that lists each, and its conclusion or status. Since I harp on you to maintain copies of everything, I would suggest you use a decent sized binder for this because the copies of the FIOA requests can get wordy and can take up a lot of space."

"Got it."

"Have a binder for statements. Have everyone you know write a statement on your behalf. I have some statements in my boxes here, but you can look online on how to write a buddy statement and send the link to your buddies. Having a binder for those statements keeps them safe and organized. Have a binder that contains your Kingdom-related traffic. This will probably be a big binder because of the amount of paperwork involved."

"Do they do anything online?"

"Timothy, I would avoid online because it doesn't give you confirmations. I used the mail system because I can track the submissions AND the Kingdom cannot say they never received anything if I have confirmation that someone signed for it. That's why I would send everything using signature receipt. Did I tell you about that yet?"

"You did, Bernie. You said to pay the extra dollar to send it signature receipt and not certified or registered mail, because anyone can sign for the envelope when sent through signature receipt."

"Good. Keep a paper copy of your submission, with the date written on the copy of when you sent the package, a screen captured image printed of the receipt through the USPS that shows the signature, date and time it was signed for, printed and stapled to it. This way you have the copy of the claim submitted, the date written on the copy, the printed receipt with the person's signature with the date and time it was signed for, and the printed tracking document at the front of the binder to keep track of all that traffic back and forth."

"Seems like a lot of paperwork."

"It is, Timothy and if you don't keep track of it, you won't remember that you submitted a claim at a certain time and have no idea if the Kingdom actually has that claim in their possession. I'm very sure that information will become very important to you...unless you want to spend the next couple decades fighting and losing? It's up to you."

"Nope, Bernie. I will lay it out as you tell me. I'm going to learn from everything you went through."

"Good lad. Depending on how your case goes, you could separate the binders even more than I explained. Whatever the Kingdom sends you, punch holes into it with a three-hole punch and keep it. You'll receive updates, decisions on your claims, delay letters, etc."

"Keep it all, Bernie?"

"What I did, to save some space, is I threw away the pages that were not important. Even with a rating decision, there's pages included that have phone numbers and generic information that the Kingdom will send to you each time they send you a decision letter. You don't need to keep those. You should keep one set that has the phone numbers and such listed on them, but you don't need to continue collecting that information. I would keep my appointment letters and when they sent me to another clinic, there was always a questionnaire they included that I needed to fill out and send back. They don't expect anyone to make a copy of that before submitting them."

"So, make a copy of those also?"

"My answer will always be yes, Timothy and I'll tell you why...it'll serve you well to maintain a copy of everything you fill out because down the road and unexpectedly, you may receive the letter that states you have to go through all of the examinations again."

"What? They rate my disabilities, then they can make me go through another exam whenever they want?"

"Not only can they, but you should expect that."

"Why? That's unfair."

"Well, Timothy, if you think about it, if you had an injury that was minor...would you expect to receive a single rating and never have to get examined again? There are many disabilities that can be fixed or get better with the right rehabilitation and exercise. However, you should also know that they use that to bully veterans also. I have gone through about five or six total evaluations."

"Were any of your disabilities considered better? Is that why they did that to you?"

"No, I believe most of those evaluations were triggered by the King's distaste for me overall. It was their way to let me know they have all that power and I had little."

"That's bullshit, Bernie...uh...sorry, I didn't mean to cuss."

"No, you're correct Timothy, it was bullshit and still is. They would use any and all excuses to show me how much control they had over me. Years back, a mate of mine had a connection inside the organization and I spoke with this person for about an hour.

I wanted to know if I was specifically targeted or if my name was flagged for any reason. The person stated they could see two of my profiles, which you shouldn't have in the system, and each profile had a different disability rating."

"I don't understand."

"I apparently had two profiles within the system and each profile showed a different rating percentage. I was also told that what they were looking at, within my profiles, was a lot of hurdles and brick walls. I quickly told this person that I do not want to know names! I didn't want names because if anything were to happen to these people, with every letter I had sent and the experiences I had, I would have been the lead suspect. I didn't need that headache."

"I get that. Why the two profiles?"

"I don't know. The advice given to me was to contact the organization and ask why there would be two rating percentages on my profiles."

"Did you do that? Did you call?"

"I couldn't."

"Why? It's clear they had you flagged or something, right?"

"Timothy, if I had called the organization and asked why I had two rating percentages and those two were different numbers, they could pinpoint the person who accessed my account to give me that information. That could have screwed that person over."

"Yeah, but Bernie, it could've busted the person who was messing with your account."

"Perhaps, but I wasn't going to throw this person under the bus in the hopes that it might help my case. I don't regret not doing that. I did write up a couple FOIA requests to gain a copy of everything that was within the organization's system. I wanted it all. I spelled it all out, too. I asked for a copy of my profiles, every profile they may have on my name, I asked for the access logs of every person, who that person was, their position, time and date they accessed my profile, the length of time in my profile, and I wanted all notes associated with each access. "

"That's a lot of information, Bernie. Did you get it?"

"I didn't. I was told that I couldn't get that information because my profile was not labeled as sensitive. I don't remember if I told you about the SPAR or not."

"We talked about a lot of stuff, Bernie. I can't remember right this moment..."

"Let's say I didn't say anything about it. SPAR is the Sensitive Patient Access Report, I think that's what it stands for, and is the report that has all of that information contained in it. The bad part was I didn't have a sensitive label on my records within the organization. After I received the letters, more than one, that stated they could not oblige my request, I found the directions online to submit the request to have my records labeled sensitive. I called and spoke with someone pretty high up and they told me that they just recently labeled my record as sensitive so I could get from that moment forward, the access information."

"What about since you got out of the military?"

"That wasn't marked as sensitive, which means they could reject my requests for that information. Nice, huh?"

"That's crooked, Bernie. How do they get away with that?"

"By being very large and greatly protected within the country. So, learn from me, as soon as you submit your claim and receive confirmation that it was received and being processed, call them and request your record and profile be given the sensitive rating. Record the conversation and request a confirmation letter that your record is now listed as sensitive. Whenever you call, make sure the first thing you do is ask the person their name. This way you know who you spoke to and what was said. I usually begin the recording before

I initiate the call and say aloud the date, time and who I'm calling. This way it's on the recording."

"Makes sense."

"Back on track now, Timothy, about organizing your information. A separate binder may come in handy to maintain anything that is off subject. For example, if you submit a letter to a politician asking for help, which I want you to think really hard before you do that, you would want to keep a copy of your letter, plus maintain anything that is sent to you regarding that letter. Usually, you'll receive a letter from their office and accompanying that would be a letter from the Kingdom to that politician. That letter usually states the same lie. It usually states how important the veteran is and how hard the King's people are working that veteran's claim."

"How come the politician doesn't do anything about the problems?"

"Long story, Timothy, but I just want you to know what to expect. They won't fight for your case, at least most of them won't. There have been a very select few who had the balls enough to fight. The rest won't. This is why I say you have to fight your fight and don't rely on others to fight it as hard as you should. Understand?"

"I understand, Bernie. I get it."

"I guess what I'm hinting at, Timothy, is you'll need a row of binders that you can organize everything better. I even had a binder for my phone discussions. There was not a lot of phone usage, but I recorded and logged each one. This allowed me to call them out a couple of times when they attempted to lie and state I never called. They moved their calls to an independent entity, which now allows them to remain disconnected when they need to, from liability. I've had to deal with the people on the phone on numerous occasions and some of those conversations were aggravating to say the least."

"You think I should have a separate binder just for calls then?"

"I do. See, if you have an appointment cancelled, they have to provide you with local options if they cannot get you in within thirty days. If it was a specialty appointment, they won't care and will offer you an appointment they know is not what you need."

"That's garbage, Bernie."

"It is. I had an appointment that was canceled and it was a specialty appointment. I got the call to reschedule from their phone center and was offered just an examination appointment. I informed the operator that my appointment is a specialty appointment specific to a location, and a generic examination with some doctor who doesn't know my case would not fulfill the requirements of what I need. She didn't care. She informed me, and this is how they do business, that she could get me another appointment with my provider for the specialty appointment in a couple months, but it was outside the thirty-day requirement. That's how they get away from their thirty-day obligation."

"They know it's corrupt!"

"They do, Timothy, they do. They don't get paid to care. They know if you want the specialty appointment, you'll agree to waiting an extended period of time to get it. It's aggravating because they keep veterans desperate for any, little, miniscule glimmer of help. So many veterans give up and just go along with it. Their time becomes a nonstop disappointment and a constant-jumping off of cliffs, following the veteran in front of them."

"Like lemmings."

"Yes sir, like lemmings. I'm sure you won't want an entire library of binders for this, but the better organized you are, the better off you will be. Things evolve, Timothy, so you may want to create your binder system differently. Having them all in one location will make it easier and more convenient to find the information. You may want to have your binders different than how I had mine, so however you can be better prepared than I was, the better overall."

"I agree, Bernie."

"If I repeat myself, Timothy, please tell me. I have a habit of forgetting what I've told you over the past, couple of days here. I do want you to know that it has been the highlight of my stay here, that you came to spend this time with me."

"Are you kidding me, Bernie? This is probably the best inter-action I've had with someone not in my family, in my entire life. I appreciate you and the information you have been willing to give me. Please don't thank me for visiting, Bernie, I need to thank you. It has been great meeting you and I only wish I was able to meet you years ago. I think we would have been great friends!"

"Timothy, I can easily call you my friend, good sir. You and your mom are two of the greatest people I've ever known, and I've been around this world for a lot longer than you have. Your mom did a fantastic job raising you, but you have done a great job making yourself into a man. You need to keep moving forward with your life and know that you are the person you are, because you have chosen to work hard and be a good person. Don't let anyone tell you any different!"

"You're a good person also, Bernie."

"I'm not that good, Timothy. I've made a lot of mistakes and I've not been the greatest person to live with or be around."

"But you had to be that type of person, right? You needed to survive and not hesitate with the job you did."

"Looking back, Timothy, I could've been a bit nicer to others, but I knew who I was and how messed up I was because of my experiences, and how I had to survive and all that. It is what it is and I am what I am, I am not going to make excuses or place the blame on others. I knew what I was and that's why I chose to avoid exposing others to me and my ways."

"Seems like a lonely life, Bernie. What about family?"

"I don't have any family, Timothy."

"Everyone has family, Bernie. Everyone has someone in their life..."

"Not everyone, Timothy, but thank you for trying. It's okay, I've made peace with the damage inside my brain and my busted body, and I face the conclusion head-on and with my boots on. I'm not scared."

"I don't know what to say to that, Bernie."

"You don't have to say anything, Timothy. Just make sure you take good care of your mother, please. She's a good woman and has been the only one who has looked out for me while I've been here and she's been ostracized because of her dedication to me and my care."

"Wait, they treat her poorly here...because of you? Why would they do that to her?"

"Because she's stirred up some problems...actually she's brought attention to some bad seeds around here and she's a tough cookie, Timothy, she stands her ground and doesn't back down. If I was decades younger..."

"She's one of the toughest women I know, Bernie. I respect my mother a great deal and have always appreciated how hard she worked to make sure life was good for the family. That's partially why I entered the Navy...I wanted to ensure I could send money home frequently to take some of the financial burdens off of her plate."

"You're a good kid, Timothy. Stay good inside, regardless how bad things can get in your future endeavors."

"I will Bernie...uh...can I ask you about your service after you finish telling me about the organization part of my process?"

"Perhaps, Timothy. If I can remember back that far...most days I cannot remember what I did an hour ago."

"Thanks Bernie."

"Did I mention the C-File?"

"I don't think so...I don't recall you saying anything about a C-File. What is it?"

"It's your file within the Kingdom. This is the file that they supposedly use to address your disability claims. I would suggest you search online for how to request your C-File and find the template that has the address and phone numbers. Once you find that, as with everything I've told you, keep a copy, put the time and date when you send the signed request, and print the screen captured receipt once your request has been signed for. This will help you because you can compare your record to what they send you."

"How will that help?"

"Your medical records in your possession may have more pages than the C-File has."

"Okay? What does that mean?"

"It means if you have pages they don't, how are they successfully assessing your disabilities? Does that make sense now?"

"Yes, it does. Holy crap! Is it they want to create so many avenues of information that no one person would know everything?"

"Sort of. Sucks, I know, but these are the important things I found over the years. This doesn't mean they haven't added processes or departments to further cause hate and discontent, so it's your responsibility to pursue these changes and keep up on it. You're going to be an agent of information."

"Secret Agent Tim!"

"I like the sound of that. With all of this I'm telling you, it can cause you problems, but with the information I have given you, you at least have a fighting chance to remain on top."

"It's going to take me weeks to organize my notes and create a template with everything listed. It's all important, but I think I may organize my notes by how the process works."

"Expound please, Timothy?"

"Well, I'm still on active duty, but the medical review board is soon approaching. I need to list it in that manner…from the Medical Board, to the results, pre-discharge actions, discharge actions, finding a VSO and submitting my claim, appeals, SPAR and stuff like that. Does that make sense, or can you think of a better way to look at it?"

"I wish I could say yes, Timothy, but I think your mind is moving in the correct direction. You can adjust your organization as things progress and I hope you never have to experience anything that causes you to add to your list, but if you don't let your guard down, I think you'll be prepared for things appropriately."

"I think so too."

"Just remain flexible and don't let all of this consume you. I lost a lot of years because of being consumed with this information and their processes. It remains such a beast for veterans because we remain divided and focused on our own cases that we don't organize together. If we were organized…boy, oh boy, would they be in trouble! Same thing goes for the politicians as well!"

"I'm sure they wouldn't want the world to see exactly how they do business in D.C., Bernie!"

"People are starting to wake up, but for most of us out here, it is too late for us to be of any benefit. We disappear without a trace, especially when you don't have family checking in."

"That sucks, Bernie, but I get what you're saying.

"As long as you keep organization at the top of your list. The more work you do now, the easier it will be. I learned the hard way, over and over again. Sad to admit, but I didn't catch on quickly…maybe because of my brain injuries, but I usually failed hard before I figured it out. I wasted a lot of time with rework, and I don't like rework or wasted time."

"I'm the same way, Bernie. I don't like to rework anything."

"Keep everything, punch holes into it and place it in its appropriate location in your binder system. Don't rely wholeheartedly on digital storage, because digital can fail. At one point, I had burnt a library of CDs, plus had an external hard drive and a paper copy. If one went down, I could rely on the other options. I used to keep a copy in a storage unit for a long time, just in case something happened to my house."

"I am planning on keeping a full copy at my mom's house. That way I know exactly where a good copy is. Plus, it gives me a reason to visit my mom!"

"I hope you don't need a reason to visit your mom, Timothy?"

"I was only kidding, Bernie. I don't need an excuse, bad joke on my part."

I didn't notice the person standing in the doorway. It was so dark that I couldn't make out the face, only the size and silhouette. Bernie had noticed the figure first and quickly made a suggestion.

"Timothy, how about we break for now? Go grab a sandwich from the cafeteria while they poke and prod me a bit. Okay?"

I was taken by surprise, but was also suspicious that the person in the doorway didn't step inside the room to ask me to leave or to poke at Bernie. I shoved my notes inside my backpack and stood up, only to notice the person was gone. I glanced at Bernie and noticed he had a sobering look on his face as he stared back.

"Sure, Bernie. I'll only be gone a couple minutes, if that's okay."

"Make it about thirty to forty minutes, okay Timothy?"

"Sure, Bernie. Not a problem at all. Need me to bring you back a bite to eat?"

"No thank you, Timothy, I think they're going to bring my lunch when they come to poke at me. Thank you for asking, though. I'll see you in about forty minutes, then. Have a good lunch."

"Thanks Bernie, I will try. See you in a bit."

I grabbed my crutches and hobbled out of the room. As I entered the hallway, I half-expected to see the nurse waiting outside the door, but no one was there. I slowly hobbled down the hallway, hoping to see which nurse fit the shadow I had spotted, but was unsuccessful. I proceeded to the cafeteria to have some lunch, but was thinking about the look Bernie gave me before I left his room. I had a sickening feeling in the pit of my stomach.

9

JOB APPLICATIONS and PREFERENCES

CHAPTER 9

JOB APPLICATIONS and PREFERENCES

I sat in the cafeteria for some time once I finished my lunch, and I thought about everything I was trying not to notice with Bernie. I was trying not to believe that he was in any trouble or that he was the target of any maltreatment. The more I thought about things, the more I caught myself checking my watch frequently, with a higher level of stress each time. Bernie was a larger-than-life character with a rough exterior, but his interior cared about other veterans. I couldn't fathom that anyone would want to hurt Bernie.

As I watched other veterans meander about with their trays of food, I noticed that many of these veterans knew each other and perked up when they spotted their friends. From what I understood, Bernie remained in his room, but I never asked my mother why. If Bernie did actually get that one bruise from the sink, or falling near the sink, that meant he walked there. I've never seen him mobile, but that's not saying much since I've been so focused on capturing every word he was saying.

The small things were flooding my thoughts as my frustration grew, and some things began to stick out. Things that I probably should have become aware of after my first couple of visits with Bernie, being: I only witnessed my mother on shift when we entered the floor, few veterans moved about on Bernie's floor while I was there, no announcements were heard while I was present at the hospital, and Bernie never complained about anything, not even in jest. I became a little angry that I hadn't notice those things before.

I finally composed myself and headed back to Bernie's room, but this time I took my time walking through his wing. I looked inside each room as I passed by, I meandered up by the nurse's station for a couple of moments, to see or hear movement of any type, and I made it to Bernie's room while keeping my ears perked for any and all sounds. Nothing different from my previous experience: zero nurses, or any other staff were around the nurse's station when I was meandering, each room I passed was either completely dark, or dark with the television on, but no patient was moving about in any of the rooms.

Bernie was sitting up in his bed, as usual, and was a million miles away when I called out to him. I loudly hobbled into his room, banging my crutches against the wall as if by accident, which got his attention. His demeanor changed as he suddenly became alert to the present moment.

"You okay, Bernie?"

"What? Yup, I'm good. How was your lunch?"

"It was okay. You were in a different place just now, Bernie. Didn't hear me call out to you? You had me worried for a second."

"No, I was just stuck in thought. That happens to us old folks...we get stuck in our own thoughts and everything else around us fails to exist. What were you saying, Timothy?"

"I was asking if you had your lunch? Did they bring you your lunch?"

"Lunch? Oh, lunch. Yes, thank you for asking. I had my lunch."

"Great."

I quickly took my normal positioning in the recliner and prepared my notebook for a flurry of writing. I tried to glance at Bernie without him seeing me, but each time I glanced up, he was locked onto me.

"What is it, son?"

"I'm sorry?"

"What is it? There's something on your mind...come on...out with it."

"No, nothing on my mind, Bernie. I was just a bit worried for you."

"Worried for me? Worried about what?"

"I don't know. You deserve to be happy and safe. I guess I just want to make sure you're both."

"Your mother takes great care of me, Timothy. You don't need to be worried about this old man!"

"My mom is not working today, Bernie."

"So?"

"So...so I haven't seen a nurse come in to take your vitals or bring you anything since I arrived."

"What's to bring? Vitals are timed and I don't need anything. If I don't need anything, they have no reason to check up on me. I know you think of this place as a hospital, Timothy, but think of this place as an apartment complex and each resident has their own lives to lead. They only get visits if it's necessary for their health or wellbeing."

"Okay."

"Are you sure you understand that, Timothy? I don't want you freaking out thinking I'm in any danger or anything. I'm doing just fine for the time I have left."

"Time left? Are you getting discharged?"

I knew it was a stupid thing to say as soon as it exited my mouth. The look Bernie gave me signaled the reality that Bernie was going to let me dangle on my words until I finally get how stupid my statement was.

"Sorry Bernie. That was stupid of me to say."

"Actually, Timothy, it was kind of funny."

"Please don't tell my mom that I said that. Please?"

"I won't...but I will tell you that your mom would have gotten a kick out hearing that. She would've laughed about it, maybe you should too?"

"It's not funny."

"It's life, son. It's something we all have to embrace at some point. If you realize you started to die the moment you were born, then your life will be a little bit easier."

"I suppose."

"Shall we talk about something different?"

"Yes, please Bernie. That would be greatly appreciated."

"It is truly funny how uncomfortable you are, son, but I won't torture you any longer. Let's talk about what I've noticed out in the workforce."

"Again...yes please, Bernie"

"I wanted to tell you what I noticed organizations were doing in the civilian sector and how I stumbled across it. First off, let me start this off by loudly expressing how strongly you should avoid filling out any voluntary profile questionnaires that talk about any disability you may or may not have."

"I thought it was illegal to ask if someone was disabled?"

"It is, Timothy, and I might have mentioned this to you previously about not volunteering your disabilities, but applying for positions you can physically perform. If I'm repeating myself, it's that important. A business can ask you if you are physically able to perform specific duties, which is not asking if you're disabled. There's a difference."

"Okay."

"I created a profile for this one organization and there was a voluntary questionnaire regarding disabilities and it clearly stated their Equal Opportunity vows and adherence, and they even had disclaimers that stated they do not bias against anyone based on disability, race, sex, etc. It's a way to get you comfortable in thinking it wouldn't matter if you volunteered that you were disabled or not. Too comfortable."

"Tricky."

"Indeed. The questionnaires differ slightly, but can all be used for the same purpose and that purpose is to be biased against someone without looking like they are biased because of their disabilities. Some questionnaires ask specific percentages of disability, while others include percentage and a list of possible disabilities."

"But if there's a list, who's to say the disability I have isn't the most minor disability on that list?"

"Indeed. The answer, Timothy, is nothing. These lists included HIV/AIDS, Diabetes, Cancer and a whole list of other horrible things. I had filled one out when I first created my profile, but never thought that it would be permanently attached to my profile. Every time I applied for a job within this organization, I would receive an email shortly after stating they hired someone else. Then, I would get the alert that that position was hiring. Since I was told someone else got the job, why would I apply again?"

"You wouldn't. No one would...I would hope."

"It took me many rejections before it clicked that I was receiving alerts for the same jobs, after I was rejected for applying earlier, but with different job numbers. This allowed them to not look as if they were being biased against me because of my disabilities. Again, some had a choice to choose if you were greater than 50%, some 70% and some 100% for disability levels."

"Why would they legitimately want to know that anyways?"

"Government jobs, Timothy, allow a disabled person to gain preference points, or percentage, for being disabled. Government positions have a 5-point preference and a 10-point preference system, depending on your percentage of disability rating. It makes it so those with a disability rating are somehow propped up, or given an advantage for a position within that organization, over and above the average, healthy person who also applied."

"How would they know? Is there an option to check a box?"

"You mean: how do they know you are a certain percentage?"

"I guess, sure."

"It's as easy as checking a box. I never thought it would be used against me. I would never have assumed that it would remain attached to my profile forever, either. I began to track the other positions I had applied for and the length of time that passed before I received the rejection email. That's what signaled to me that something was happening. A pattern."

"Holy crap, Bernie. That's such BS!"

"Remember this, Timothy, have resumes ready for government positions and resumes prepared for civilian positions. This way you can safely type in your preference points at the bottom of the government resumes without it affecting your chances at a civilian business. There's also another form that you should have that has your preference information on it. I think it was labeled SF-15 or something."

"Is it online?"

"It is. You could download the form and fill it out, then sign it and scan it into a PDF. That'll make it easier to attach to government positions."

"Are you sure an organization would tempt using their disability questionnaire to bias against someone who is disabled?"

"Not at all, but can tell you that things can only occur so many times, in the same manner, before one would need to look at it closer and conclude that was the sum of suspicion. To be safe, you should avoid filling it out. It's no one's business and as long as you apply for the jobs you know you can do, then you don't have admit you're disabled at all."

"I'd do that anyways, Bernie. I'm a terrible liar and I am not willing to lie just to get my foot in the door."

"Good lad. Some organizations have you create a profile so you can come back and apply for other positions, while others don't. Anything you enter into a profile that you create could remain on that profile indefinitely. With this one organization, when I logged into my profile, I needed to update some information and the first thing I did was uncheck the disability blocks."

"Did it change anything?"

"I received immediate an response asking if I could do an online interview."

"That's good."

"It was, and it was fast. I did the online interview and it went well. Within a day I was asked to come in and do a face-to-face with the manager, which I did."

"Did you get the job?"

"I did not, for a couple of reasons. The important thing to remember is to avoid admitting information that is not directly pertinent to the subject at hand, which should be to apply for that job."

"It sucks to have to worry about that also."

"It does, but knowing now what you know, you should be in a good position to avoid the garbage that I faced. Remember, they can ask you if you are physically able to perform the job, and actually, most of the jobs that came out in the past years, have listed the physical requirements of the job. This allowed the businesses to be clear and up front about what the demands of the positions are."

"What if someone applied who was disabled, but didn't admit to that when they applied?"

"They can be fired."

"But that's being biased against the disabled, right?"

"No, it's not, Timothy. It's the applicant's responsibility to read over the requirement to identify if they have the required qualifications and abilities. Lying on an application is grounds to be fired and no amount of screaming about it being unfair or biased against you, would change that."

"That does make sense, though. I'm sure some lie on their resume and say they can do the job, just to get inside. I guess they figure they can figure it out as they go along, or fake their way through things long enough to figure it out. What is the phrase? Fake it until you make it? Yeah, that's never a good thing."

"I couldn't have said it any better, young man. You really are a bright boy. Your mom is a great woman to have raised such a smart son!"

"She may argue that point, Bernie."

"She probably would. She likes to argue about things."

"What else have you experienced in the workforce, Bernie? Things I should know about?"

"Well, Timothy, you should know that civilians rarely understand what anyone in the military did or does. You should probably avoid making slang comments as if you were still in the military, because the civilians will just look at you cross-eyed. Remember when I told you that it doesn't matter where you go, there will be good people and bad people?"

"I remember."

"Well, you're going to find that some civilians are not entirely focused on completing jobs as quickly as you are. They may seem a little lazy or unconcerned about prioritizing their work. This can be aggravating to say the least, to anyone who spent any time in the military getting their work done quickly and without rework. Now, I'm sure you experienced service members who were lazy and rode the coattails of those who were good workers. Some of them probably skated right on through their careers, while some of the hard-working soldiers suffered."

"I've seen that."

"Same goes in the civilian sector. There are those who will brag about being a veteran and the civilians around them think they are a warrior, a war hero and a great worker...yet they turn out to be duds. Same goes in Washington D.C., Timothy. Politicians may use their status in the military as a way to get elected because they know the average person has no idea what they did. There are SEALs who run for office who I wouldn't trust with a rubber band, yet the civilian populace love to goo and gaa over them because they wore a bright, shiny gold trident on their chest."

"I have to say, Bernie, I would probably give false respect to a SEAL because they were a SEAL...before speaking with you, that is."

"Most would, Timothy, but who's to say that any of these representatives were honorable, or served honorably? Just because someone enters into the military, spends a couple years and then gets out, it doesn't make them a hero, nor a morally-driven person. It doesn't even make them a good person. Some might say I wasn't a good person."

"You weren't a good person, Bernie?"

"I said some might say I wasn't a good person. I was a good person, but people didn't like that I would call them out for their BS! Many times, I got threatened by some of higher ranking, who quickly learned that I didn't care too much about following the chain of command when someone threatened me, or tried to use rank to bully me. I took it for action. Will some call that nice? Probably not...okay, definitely not, but I've never made excuses for who I was, and I'm not going to start now."

"Nor should you have to, Bernie. But, were you a good person?"

"If you mean: did I make ethical decisions? Did I perform my job morally and while trying not to do harm to others, then I would say yes! I went out of my way to do good things for others. I stood up for those I needed to protect, because they were new to the community or they didn't know what they were doing, and I never made excuses for doing so. I was the guy that would go up to the Commander's office, knock on his door and explain how his view of certain things was incorrect."

"What? To the Commanding officer?"

"Yes sir, Timothy. Do you know why?"

"Because you weren't a nice guy? Only kidding."

"No, I did that because we had a great Commanding Officer who had an open-door policy and stated numerous times that if any of us thought he was full of garbage, then come up to his office and tell him so, but they better have a solution in mind. He wanted to begin a dialogue."

"And you took him up on that offer?"

"Many times, actually. I've walked in and told him straight that he was jacked in his statement to the Team, and he would point to the chair next to his desk and invite me to sit and discuss it, which I always did. I knew my role. I knew what I knew and I knew few had the guts to approach this CO the way I would."

"Was he scary? Most COs are scary because of rank or attitude."

"He was a very large man from the secret squirrel command and was very intimidating. I wasn't intimidated, but I loved that he used his stature and everyone's fear to create more fear. I thought it was funny. After discussing solutions with him, he would call the department that could help make changes and bring them into his office. They would see me sitting there and be confused as to why I was there. The CO would always point and ask the department head if they knew who I was and they would say they did."

"So, the CO would ask if they knew you?"

"Yup. Once that was established, the CO would then open the floor up for me to discuss my solutions to whatever problem brought me into his office. It was usually met with a positive reaction and discussion, but a couple times it didn't. Oh well."

"Is that why some didn't like you, Bernie?"

"Could've been. I believed in working hard and playing hard. I always looked out for my men and wouldn't let anyone else come after them. I suppose that's why I never went above a certain rank. I always believed it was for the greater good. I didn't really fit into any one group, sort of a loner most of the time, but I wouldn't play the game and eventually, if you want to gain rank, then you have to play the game."

"I try to remain low-key. I do what I'm told and try to stay away from the drama, but drama usually finds me eventually."

"Timothy, you're a good man and you know to distance yourself from the drama. Don't feel bad that you don't play the game like others do. You need to be yourself and like who you are. It took me a long time to figure that one out. You're still young and if you learn that now, you will have a glorious life ahead of you."

"Bernie, what would you do differently? If you had to do it all over again, what would you do different."

"I don't think I would necessarily do anything different. I experienced a lot of things in my life and made a lot of brothers, who I loved just as much, if not more, than my real family."

"Do you have any family still around?"

"You mean are they still alive, Timothy?"

"I guess, sure."

"I don't know, but I don't think so. I disconnected from them a long time ago and have been on my own since. I don't regret anything, which is something you should embrace now...don't live with regrets! Fail as often as you need to, to achieve your goals successfully! Failure shouldn't be feared...it should teach you how NOT to do things."

"I think it was Edison that was asked how many times he failed in his light bulb designs and he said he didn't fail, he found 99 ways how not to make a lightbulb and one way to make a successful one...or something like that."

"That's a good example. The mysteries of the world remain mysteries to those who never pursue the possibilities. If you want to start your own business, start it. Do your legwork and research first, but there are a lot of opportunities out there for veterans who want to start their own business. If you find your local Small Business Association and sit down and speak with them, they can help you out."

"The Small Business Association? Does it have to be a specific business?"

"It doesn't. They are there for help and to guide you towards being successful. They are pretty helpful and willing to take the time to guide you the right way. For veterans, there are programs that will help you out further. Don't shy away from those programs and don't think that you don't deserve them, and especially avoid thinking that if you decide to ask them for help, that you are somehow taking away opportunities from other veterans."

"Does that happen?"

"Do veterans think that way? Sure. They also think to themselves that they shouldn't file a claim for their disabilities because that would take money away from other, more needy or more deserving veterans. That thought is insane!"

"I don't get it."

"Timothy, some veterans leave the military feeling half-way decent. In other words, they don't hurt and are able to still walk and talk with relative ease. They leave the military and say to themselves that they feel good now, so they won't file a claim because they don't want to take that money away from veterans who need their disabilities granted. That's a foolish thought. What happens is, those same veterans begin to hurt ten or fifteen years down the road and suddenly they wished they had dropped a claim when they were discharged or retired. Then, they finally decide to file a claim."

"That's a big time gap."

"It is and it will be nothing short of an uphill battle for them! They would need to prove those disabilities affected them the entire time. How many of those veterans do you think visit the doctor's office during that time? None, the answer is none. If they don't feel bad, they won't go. The military teaches each member to avoid going to medical unless they are absolutely in dire need of medical attention. I'm sure the military hasn't changed that much in that avenue."

"No, it's pretty much the same. It's looked down upon to repeatedly run to medical. It also impacts your ability to do your job."

"Then it's exactly the same as when I was in. We had those we called sickbay commandos, who would conveniently rush to medical to avoid difficult periods of training, or what have you. They're everywhere. Can't avoid them. However, it is silly to think you would impact anyone's money if you were to submit a claim. The more time you delay in submitting your claim, the easier it is for the Kingdom to deny you."

"That makes sense, Bernie."

"For those people who wait a long time to submit, the Kingdom will request proof of ongoing disabilities through civilian medical documentation, or military if that person retired. They want to see that a certain problem has been bothersome and if the veteran truly had a problem, there should be an ongoing paper trail of documentation from medical professionals. Most veterans won't have that proof."

"But you know of vets who thought they would be taking money away from veterans who deserve it more, if they submitted a claim?"

"Of course I have. I had mates who thought that way...then a decade later they contacted me and begged for help. They also complained about how unfair it was that their claims were immediately denied. The King loves to use the excuse that you, the veteran, didn't go to medical repeatedly for the same injury or pain, so to them that means the pain is not chronic, nor severe enough to deserve a rating."

"Wait, what? But they should know that we don't run to medical all of the time, for the reasons you just mentioned. We don't want to be labeled as a sickbay commando! That's ridiculous!"

"They do know that, but they won't tell you that they know that. They won't admit they know service members avoid running to medical constantly. This gives them yet another arrow in their quiver. They get money each year for the estimated veterans that depart the service."

"They do? How do they know how many would get out?"

"They either have an estimation from the government and they add a specific percentage to that to account for a plus or minus amount, or they look at trends and estimate that way. Either way, they budget it in. So, if you don't submit a claim when you are discharged, the money they estimated for you, specifically, goes into their pockets...or as they like to say...it goes into other programs! Teehee."

"They keep the money?"

"They are a government entity so they have to show they spent all of that money for that fiscal year, or they won't get the same amount the following fiscal year. They shouldn't get more money, over the amount they previously requested, if they didn't spend it all. It's exactly the same in the military. Spend it all before the next fiscal year shows up, of you will lose out."

"I've heard about that but I wasn't part of any spending information."

"Well, know that if you don't file a claim immediately, then the money they allocated for your possible disability claim, will go into their coffers to spend as they see fit, which can also mean very large bonuses."

"They can keep the money they allocated for me?"

"If you don't claim anything, sure. They don't technically keep that money...more like they can deposit it into an account that can grow interest. This way, if and when you decide to submit a claim and can prove it was service connected, they have money available to pay as such. That's the hook, though, Timothy. They avoid admitting service connection for some veterans. Decades, young man, decades I had to fight for my disability ratings and everything I claimed was inside my records. The proof was there, yet they constantly retitled disabilities, combined disabilities together that were not supposed to be combined, and so forth."

"Underhanded! Unethical, Bernie. How do they constantly get away with that?"

"Because they have many avenues to move things around. They have many politicians who can provide extra special support when needed, and in return may receive a nice donation into their accounts for that help and support. It's such a big system, they have been smart enough to avoid being exposed fully. If someone gets caught doing something that the public deems unacceptable, they move the person responsible to avoid litigation and exposure."

"Sneaky."

"Indeed. They got exposed in one state for deliberately messing with the number of veterans they were servicing, to make it look like they were processing their requests in a quick manner, but when they were exposed, it was proven that that specific location was deliberately increasing the numbers to prove they needed more funding, but also to make it appear as if they were taking care of those veterans."

"They should get fired!"

"I agree, but many times they don't. The guilty person would be moved to another state and placed into another position, which then endangers the lives of those veterans."

"How is it they haven't gotten caught yet?"

"They move that person in the shadows. There was a doctor who was showing up inebriated and had sexually harassed some veteran patients. The staff knew and said nothing. A veteran finally said something and turned him in and then, only then, did the King do something about it. Another surgeon was conducting surgeries he wasn't qualified to perform. Once he was exposed, after a long time after he was exposed, he was dealt with."

"But they don't fire them? I don't understand how they are allowed to remain within the system that caters to veterans?"

"The joke back in the day was that these doctors who come to the Kingdom, do so because they can't be sued for malpractice. That rule changed and suddenly many doctors decided to leave the Kingdom. Odd timing. I watched the King quickly retire executives when the exposure of what they did remained in the media spotlight. The King couldn't hide it or relocate them, so they allowed them to quickly retire with a fat severance check, and then they disappeared from the public eye!"

"And the politicians do nothing about that? That's crazy!"

"One of these so-called ethical people was a lawyer who had his own veteran nonprofit organization. He was exposed for pocketing a lot of money and when his organization was looked into, it was found that this lawyer, who was already getting a lot of money from the Kingdom as their counsel, had only spent a couple thousand, out of the many hundreds of thousands of dollars donated, on veterans. Once it hit the media focus, he was suddenly fast-tracked into retirement by the King."

"I suppose that means he would be receiving a monthly check for his retirement then?"

"That's exactly what it means, Timothy. Plus, their severance payment, which is probably close to six figures, depending on their position and status. It's sickening. That is what motivated me to keep an eye on the veteran Bills in Congress and the problems that the veteran community suffers from."

"There seems to be a lot of issues within the veteran community from some of the headlines I've seen. Like homelessness and suicide."

"Correct, but if you see that, then you can expect those problems to be much, much worse because it has to get noticed and be very bad in public opinion, before anyone with do anything. Think about this: how bad is veteran suicide?"

"Pretty bad...I think the last word I heard was, like 15 a day?"

"Do you remember what the number was for many years?"

"Twenty-two? Is that right?"

"It was twenty-two, but the actual number was higher."

"How much higher?"

"Higher, Timothy. The media avoided publishing any numbers because it would look bad on the politicians, who knew it was bad, yet did nothing about it. Oh, they'll brag about the Bills they propose, but they won't admit that those Bills were proposed deliberately to look the part, but they weren't going to do anything with those Bills. Those Bills were not going to move forward, but they will brag up a storm about how many Bills they proposed and how they love their veterans within the community!"

"How do they get re-elected, then?"

"People do not do their research, Timothy. Who has time anymore? What would you rather do, listen and believe what the politician who you voted for, says, or listen and spend your time researching to see if they're actually moving their Bills forward?"

"Probably listen and believe."

"Exactly. The media does shady reporting all the time. They state one thing, then they get fact-checked and exposed for outright lies, so the media will go back to their original article and type a retraction article, or a small correctional blurb."

"That's appropriate, Bernie."

"You are correct, good sir! So let me ask you this: How many articles, of the ones you have read, have you gone back and read them over again, just to see if they had to write a retraction on anything?"

"Uh...none! I wouldn't think I needed to...I would think that the article was written after precise research and fact checking was accomplished...oh crap!"

"There you have it! Very few will ever go back and reread an article they've already read. Why bother? You already read it once! This is how propaganda is pushed like a virus! So many people are too busy to do any level of research, and then they act like hurt puppies when the problem is at their doorstep and affects them directly. Then they want action!"

"Typical! I had worked with people like that before. It sucks."

"It does, but it's reality. This is why you need to research things and keep on top of the information that is important to you. Do your research and then gain your opinion, but the politicians and the VSOs were not going to allow themselves to look bad in the eyes of the population. That's their lemmings they rely on to vote them into office again, or to donate lots of money to. If VSOs are shown to pocket money, instead of actually helping veterans, they would lose a lot of money. They don't want to lose a lot of money, or any money for that matter. Same goes with politicians."

"Our whole world is corrupt!"

"I'd like to think that there are some good people out there, who are impervious to corruption. Unfortunately, the bad outweighs the good. Veteran suicide is a very real problem, and the majority of those veterans who have been committing suicide were veterans from wars that ended many, many decades ago. These are the veterans that should have already been taken care of, yet are not."

"We lose a lot of vets and that's not right, especially when we watch our politicians protect illegals and violent felons and such."

"I agree, Timothy, I agree. We have homeless veterans, we have abused veterans who are in veteran facilities, veterans who die from inept care providers, we have retaliation happening against veterans, and we have veterans who cannot get jobs because, as I stated before, they get biased against because they admit they have disabilities. All of these issues and nothing moves in the direction to help them."

"Yet, the politicians will keep proposing Bills, over and over again?"

"Correct."

"No one notices the Bills they propose are the same ones that were proposed in the past Congresses?"

"Evidently not. I notice that trend, but politicians won't answer to letters that point this out, because they may owe the King favors or the other politicians own them. Sad to say, but that's the reality we currently live in."

"That's absurd, Bernie!"

"Which part?"

"All of it! I don't understand why or how they are allowed to continue doing this to people? It's aggravating and disheartening!"

"It is, and that is why I chose to track what these people were doing. It's all in what you do, that matters. Take care of yourself, take care of your mother and do the right thing with your life and your decisions. There are bad things around us, we won't change that, but we can impact those we interact with by engaging in positive actions that help, not hurt people. Live a good life for you and no one else, and happiness with follow."

"Are you happy, Bernie?"

"Son, I don't know what happy is. I thought I did once, but was wrong. My life was filled with adventure, danger, excitement and mystery. I survived when I shouldn't have and was blessed enough to experience the world. The world awaits you, son, and it's yours for the picking. You just have to move forward. There will be bad people who try to block your path, as the Kingdom had blocked mine many times, but it's those moments and what you do to rise up from them, that will help your direction become smoother."

"You sound like a philosopher, Bernie. I wish I gotten to know you a long time ago."

"Me too, son. The positive part of this is: we have met and we have become friends. Regardless what happens around us, no one can take that away. The information I leave you with should give you enough direction to do your own searching and collecting of information. This way you can use your notes as a guide to find the correct information or validate the information I gave you. As I mentioned earlier, my information may be outdated because tactics change, especially when too much of what the Kingdom does, gets exposed."

"I wish I didn't have to leave tomorrow."

"Me too, Timothy, but you have to go back and begin your life. I know you feel bad that you couldn't continue with what you were doing, but you're not done yet and your fate hasn't been sealed. Keep a positive attitude and organize your notes and things should be smooth sailing for you."

"I hope so. Are you going to be okay here, Bernie?"

"Okay where? Here? Of course...I survived situations that should have ended me, so this place ain't going to get the drop on this old dog. I am always a step ahead."

"If I give you my number and address, will you reach out if you need anything?"

"Of course, Timothy. Thank you."

"Bernie?"

"Yes sir?"

"I'm serious! If you need anything! Anything at all, please call me and I'll make sure you have what you need. Plus, you could always call if you need to complain about my mom! I won't mind!"

"That's kind of you, son. I will remember that and I will remember you, Timothy. No matter what news you get, no matter how dark your world becomes, at the other end is always the light, son. Can you remember that for me, please?"

"Remember that? That's sort of cryptic, Bernie. What're you trying to tell me?"

"Nothing other than what I have told you up to this point. Sometimes we get news that seems like our world has collapsed around us and the dust fills the air and obstructs our view of what could be. Eventually, even the dust settles given enough time. I just want you to be prepared for everything."

"Definitely sound like a philosopher, Bernie. I will remember that, it's good advice. I'm sure I will have a difficult time with everything as it steamrolls towards me, but with my notes, I think I am in a much better position to defend, more than most service members have in hand when placed into this position. You've been a huge help and a great influence for me and I want to thank you for that, Bernie."

"It's been my pleasure, son. Make sure you call your mother when you get back safely. I'll be waiting for her to update me on that. As long as I'm here, you can reach out if you have any more questions. I gave you a mountain of information and I don't expect you to have absorbed it all, so reach out if you need me to answer any questions that you think up."

"I will, Bernie. Thanks again, I am eternally grateful for your help. I truly mean that."

"I know son and you're welcome."

I took that to mean Bernie was getting tired and didn't want to delay me going home. I spent another hour with Bernie talking about his experiences while he was on active duty and was blown away by some of the things he accomplished. I was in awe, but could tell that he was telling me this information to give me some insight into what made him who he became, and not because he wanted me to celebrate his guts. It was a sad moment for me to pack my bag for the final time and leave Bernie in his bed, but we both knew it was inevitable and we both embraced it bravely.

I had my backpack slung and hobbled over to Bernie's bed, trying not to slam my crutches into his side rails or cords, where I stood with my hand extended. Bernie grasped my hand and gave a strong, firm handshake. It was then I noticed how bloodshot his eyes were and how the dark bags hung from underneath them. I acted like I didn't notice how exhausted he looked and, for the final time, thanked him for everything he did for me.

"Timothy, amongst the long list of things I wanted to make sure you heard and understood from me, an important factor to this is this: if there's a chance that you could help another person avoid a lifetime of pain and aggravation by repeating what I've told you or what you learn through your experience...pay it forward and help them out. Can you do that, son?"

"Of course, Bernie. That's a no-brainer!"

"Good lad. Now you take off and spend the evening with your mother. Don't forget to let her know when you arrive back at your duty station, so she can let me know. Okay?'

"Yes sir, I will do that, Bernie. I will do that."

Bernie nodded and patted my hand. I turned and crutch-walked my way to his doorway, where I stopped and turned around. I gave Bernie a final head nod and removed my hand for a moment to give a short wave. Bernie nodded and waved back, turned his head and slowly settled back onto his pillow. I could see Bernie's silhouette from the door from the lights behind him on the wall and the machines at his bedside. I didn't realize it then, but this was going to be the last time I would see him alive. I would have remained longer or studied his face before I left, to weld his image into my memory, but I didn't.

I left the hospital and rode back to my mother's house, where I packed my things and sat with my mother for hours over dinner and coffee. The entire time, my mother spent her energy convincing me that she would look after Bernie and ensure nothing bad happens to him. I found this odd, because I knew how much my mother cared about her patients, especially Bernie. It wasn't until much later that I would figure out why she spent that energy.

My mother's focus was to take care of the veterans at her hospital. She took her position very seriously. She would go out of her way to give these veterans the care they deserved and the connection they yearned for, especially since most of those veterans didn't have any family, or any visitors other than old friends. With me being away, she shifted her love and attention towards a cause she felt so strongly for.

10

THE ENDING

CHAPTER 10
THE ENDING

A week after I arrived back at my duty station, I received the news that Bernie had passed away. I was stunned at the news because his memory was so fresh in my brain. I just met this man, just got to know this war hero, and just became friends with him. Now, I will never get to speak with him again and that, in itself, was a travesty. I should be thankful for the time I did get to have with him and cherish the information he entrusted me with, because that would be honoring his memory.

My mother was understandably upset at his passing, as she was closer to him than the other veterans, but she seemed extremely affected by his death. She wouldn't provide me any details other than she was with him in his final moments and said he smiled when she mentioned my name. That helped heal my internal ache of sadness. For a fleeting moment I thought that perhaps my mother was saying that to make me feel better, but that thought disappeared as quickly as it surfaced, because my mother never sugar-coated anything. This is part of what made her a great woman.

I followed through with Bernie's instructions and, although I had to experience much of what he mentioned, his preparation was crucial for me to avoid the same pitfalls that he spent so many decades trying to crawl out from underneath. At my medical review board, the deciding members didn't truly listen to my answers to the questions they asked, it seemed more like going through the motions. My appointed military attorney did not show any level of intestinal fortitude, because the board definitely outranked my attorney, and it showed.

I did learn some valuable lessons through all of this and it feels like the right thing to do, to pass on this information for those who may need to have a starting point in their lives, to pay if forward and help them out. Bernie would have wanted me to do that for them, as he did for me. That is exactly what I did. At the time, the hours I spent with Bernie, frantically writing to capture every word he said, I didn't take him as entirely serious at first. As time progressed, he didn't change his delivery, his attention to detail despite having brain issues, and his desire to try and save me from the process and the pitfalls, which made me more invested.

My life will always be filled with pains, aches and stiff muscles. I'll forever feel the weather change before others, and be miserable while putting on a fake, positive face so others don't keep asking me if I'm all right. I didn't wallow in self-pity, I didn't blame the system or give up. I pulled up my pants and I delivered an organized defense for my own case. Bernie did a great job in hyping up my self-reliance and inner fight for my case. As Bernie would say: "No one will fight as hard as you should for your own case!" and that's true.

I did make 3 copies of every medical document I had in my records. I reached out to the clinics and hospitals I had visited during my time in the military and requested a copy of my records, which they gave me on a CD. I burnt three copies of those discs, but I also printed every page they had on those discs. I went to the local Office Max and did it myself, to ensure every page scanned and printed legibly. I wasn't going to trust anyone else to do that, because my thinking was this: If I walked into Office Max and asked them to make a copy of each page, they will scan and print each page, but that doesn't mean they check each page to ensure any can be read.

Call me untrusting, but I took stock in my responsibilities with my own case and was moving ahead with my preparations. I checked out my service record and did the same thing, as well as my dental record. I only made two copies of those since I was pretty sure I wasn't going to need more than two. The important documents were my medical records. I placed my service record into a small binder, and did the same with my dental record. With those printed, punched and placed into a binder, I moved onto my medical records.

I made one copy of my medical records and documentation from the hospitals and clinics I had visited while on active duty. Once I had a readable copy of everything, I numbered each page, to include the two-sided pages. Once that was done, I made two more copies of those pages. I did this because tabbing came next. If there was ever a question about any of my disabilities, I could refer the organization to a specifically-numbered page within the medical documentation. Bernie would be proud, he didn't recommend that, but after his tutelage I felt empowered to find more ways to ensure I was better prepared.

I used colored post-it notes that were small and perfect size to include numerous tabs on any page that had more than one dis-ability or problem on it. On some pages, I may have had 3 different color tabs, due to the extensive damage from the trauma. I could quickly jump from one green tab to the next, without needing to read the small print. All green tabs were for my back, which included my lower, mid and upper back and neck. I did the same with the other colors. I think Bernie would have been surprised about that organizational tactic.

I spent a lot of time doing deep research into the most popular VSOs, the local and the VSOs through the state, and I spent a lot of time talking and listening to other veterans while at the hospital. I was happily surprised at how passionate each person was, whether pro-VSO or anti-VSO, they all were passionate about me not getting tied with them or me signing with the ones they considered good for veterans. I watched the Committee hearings where the VSOs made their speeches and I watched their body language while listening to what they considered to be the big problems within the veteran community.

After spending numerous hours watching something that I feel wasted hours of my life, it frustrated me to hear some of the same problems they screamed about the year prior during a Committee hearing. My aggravation was in the thought that most who watch these hearings, probably don't watch every hearing over the years to witness how repetitive their faux passion was. My question would be: if you're this passionate, why is there a need for a repeat of the previous year's concerns? If you advocate for veterans, then advocate and make that change happen.

My experience reinforced what Bernie had warned me about. Did I go with a VSO, or did I go with a law firm? I don't think Bernie would want me to answer that question, because he would have wanted me to tell anyone listening to do their own research on possible support. This way no one would choose a certain VSO just because I had a good experience with them. If I did that and someone signed on with the same organization in their area and was screwed over, they would cry the victim and point the blame this way. Like I said, Bernie taught me well.

Did I do my research on law firms? I sure did. I made sure to search and verify the list of law firms and lawyers on the accepted list. Their responses to my inquiries were close when it came to the percentages. Most stated they would get 30% of any retroactive check for every claim they submit on my behalf, and if I fired them or they fired me, I would have to pay for the hours they had invested in my caseload. Par for the course. Did I go with a law firm? Refer to my previous answer, as it has not changed.

I became proficient in Microsoft Excel and after I went through my entire records and paperwork printed from the CDs, I created tracking documents so I could track the progress of each disability. Being organized made it so much easier to use columns to update the actions taken. In some cases, the disabilities were combined, but I was able to add notes so I would remember the actions taken with each, specific disability. I noted if service connection was granted for the disability and its rating percentage. Just by scrolling down and/or across the document, I could see the status of every disability.

I did as Bernie suggested, I submitted a letter (yes, I kept a copy of it) and called to request my record gain the label of sensitive. I wanted to make sure I was able to FOIA request the SPAR for my records, if and when it was necessary. I made it a point to highlight that note from Bernie, because I could tell he regretted not knowing that information in the very beginning of his discharge experience.

My experience with my college fund was boring. I was able to continue my education using the Bill I invested in when I joined the military. I noticed that there were programs that provided more funding, if I wanted to pursue graduate-level degrees. The information exists, it's out there, and at everyone's fingertips. One just has to search and read. I would suggest maintaining a logbook of sites you visit and rate those sites. This will help avoid wasting time by accidentally rereading previously-reviewed sites and information. It also allows you to go straight to the sites you trust the most.

It was helpful that Bernie had stressed I wouldn't be taking money away from other veterans, who I would think deserves it more than myself, because I would have been one of those people who thought that, and I would have gone without. The money is there, for every veteran that gets discharged, one just needs to ask.

Although Bernie couldn't expound on internet-related items, I figured I would add onto what he had told me and try to help others with what I found. Many, and I cannot stress this strongly enough, MANY social media groups are filled with veterans who are desperate for information, but they cannot view that group through independent eyes and clear thoughts. What could that possibly mean? It means some of those sites are merely to gain high numbers of followers and boost their egos and give them the control they are so desperately needing.

Not all of those groups were poorly organized, but they are touchy about being called out for misinformation. A group in fear of being exposed will quickly block the person, or persons, they feel are their biggest threat. I made the mistake of calling them out because they were giving poor information and causing veterans to shift focus from their concern and their focus, towards a direction that was only going to get them lost within the endless pages of information. This made many veterans question the validity of that group.

I had noticed, previously, that the group admins were claiming to be retired staff at the big organization, or VSOs for other organizations, and that raised some red flags for me. I read through many of their discussion topics, their quickness to remove comments and commentors from a discussion thread, and finally the misinformation they were deliberately shouting out to the masses, while ignoring the replies that were questioning the validity of that information. This made me aggravated, to say the least, because that was deliberately deceiving the veterans with questions.

It was very difficult to tell the groups that had valid information from the groups that were deliberately deceiving their audience. A big clue was the demands from the admin, to never question their statements. Some good information was sprinkled about, but it seemed that whenever I spotted disinformation and deliberate focus-change, it came from someone who claimed to know it all, and had worked for the big organization for many, many years. It became clear that there was a good chance that infiltration could occur.

I stopped reviewing these groups on social media when I started to see more group administrators screaming "Trust Me" and/or "You shouldn't do that because you'll never win". These were statements that cause veterans who are at their last wits, to believe their claims shouldn't be pursued, despite suffering with the disabilities they were trying to claim. That's part of the 'Big game' though, to cause aggravation and irritation until the veteran finally gives up. Bernie taught me to never give up.

I am thankful to this day that Bernie became a part of my life, even for only a couple of days. He reminded me that information can easily die with its professor, unless that information is passed along. Will it all work? No, that's evolution. Tactics change, rules change, wording is changed, ratings are changed, and so on. It's apparent that the information can be helpful to so many, if only they pursued that information. Not knowing it exists causes many veterans to quit overall, and suffer in silence.

There are some who will be lucky enough to smoothly move through their claims and overall experience, while others will have to battle. Screaming and crying about being a victim will not change anything, except weigh that person down and cause depression, and the overall feeling of being a failure. With the information that exists out there in the universe, it's the effort, a little effort and the desire to not become the victim, that is needed by each. That little effort can change their whole world. It did mine.

The kingdom will only grow larger and its troops greater against the community. If the people refuse to organize to become a much stronger Army...the kingdom will continue to grow larger and more abusive. It's once the Armies begin to organize and become more aware, that the king and its soldiers will begin to realize they have been exposed.

I know Bernie would be happy that others could hear his story and words of advice. My mother was reassigned to another floor the day before they found Bernie deceased. She tried to sound brave, but I could tell that the reassignment and coincidental death of her friend and mine, forced her to believe she had failed Bernie. Something bigger was happening and I believe my mother may have had an idea what that something was, but she couldn't prove it.

I suppose if there was to be a moral to the story, it would be many: pick the brains of the older veterans, as they have many more years of experience; take a little time to get to know a veteran at the care facility, or hospital and listen to their stories because many have no family left; organize your information and keep organizing it until you have everything the way it should be; never give up; and finally: live your life and don't let things consume your time.

As Bernie would say: We are only on this Earth for a short time...live it to the fullest, be a good person and fight for what you believe in!

I miss Bernie and I wish I had more time to hear more about his history. I'll never forget him, and I refuse to give up on anything that is important to me. I'm sure it would be Bernie's wish for all veterans also.

9 798985 702002